HISTORIC HIGHWAYS OF AMERICA

VOLUME 13

FOUR ROUTES TO THE WEST

[*An interesting view of Juniata River near Newport, Pennsylvania; the abandoned canal; the old line of the Pennsylvania Railway; and the new line of the railway*]

The Great American Canals

BY

ARCHER BUTLER HULBERT

With Maps and Illustrations

Volume I

The Chesapeake and Ohio Canal

and

The Pennsylvania Canal

AMS PRESS
NEW YORK

Reprinted from the edition of 1902-05, Cleveland
First AMS EDITION published 1971
Manufactured in the United States of America

International Standard Book Number:
COMPLETE SET   0-404-03420-9
Volume Thirteen   0-404-03433-0

Library of Congress Number: 70-147112

AMS PRESS INC.
NEW YORK, N.Y. 10003

CONTENTS

ILLUSTRATIONS

PREFACE

CANALS played an important part in the later pioneer movement westward in America. Two monographs of this series, therefore, are devoted to the rise and building of three great canal routes westward, the Chesapeake and Ohio, the Pennsylvania, and the Erie canals.

The present volume is devoted to the Potomac Company and its successor, the Chesapeake and Ohio Canal Company, and finally the Pennsylvania Canal. In each case the birth and development of the two great railroad routes which follow these canals, the Baltimore and Ohio and the Pennsylvania Railways, is also sketched.

The history of the Chesapeake and Ohio Canal is contained in the reports of the company; Hon. Theodore B. Klein has written a very interesting account of the Pennsylvania canals, *The Canals of Penn-*

sylvania, and the System of Internal Improvements of the Commonwealth; Mr. William Bender Wilson has included a fine sketch of the Allegheny Portage Railway in his *History of the Pennsylvania Railroad Company.* To both of these the author is indebted for advice and assistance.

A. B. H.

MARIETTA, OHIO, March 1, 1904.

The Great American Canals

Volume I

The Chesapeake and Ohio Canal

and

The Pennsylvania Canal

CHAPTER I

CANALS are of two classes, those admitting large and those admitting small craft; the former are technically known as ship canals, the latter, barge canals. It is, of course, the barge canal, in its relation to the western movement of the American people, that agrees in all essentials with our present study of Historic Highways and which should be considered in any study of the subject.

The subject of fast and safe transportation of freight has become so commonplace in our day of railways that it is with difficulty that we catch any true idea of the economic importance to our forefathers of the invention and general use of such an ordinary thing as a good wagon. The meaning of the successful opening of a great canal, such as the Erie Canal, can hardly be understood unless one has known nothing of the problem of transportation save as represented by the pack-saddle and " Conestoga " wagon. When looked at from such a standpoint, the lock canal is at once seen to be one of the grandest inventions of any age; it was every whit as far ahead of any system of transportation when it was discovered as the railway is in advance of the best canal today.

From this point of view — that of the comparative value of this method of moving freight over any method known before it — it must seem most inexplicable that the lock canal was the invention of moderns. The simple canal lock, with all its immense benefits, escaped the notice of the builders of pyramids or the " Hanging Gardens," of the Parthenon and of the engineers of the Cloaca Maxima. Egypt, Babylon, Persia, Greece, Rome, with all their vast needs in the way of handling heavy freight, never invented the simple hydraulic lock. And this is the more astonishing because they dug great canals; the Royal Canal of Babylon was twice as long as our American Erie Canal; the Fossa Mariana, from the Rhône to the

Gulf of Stomalenine (102 B. C.), the Emperor Claudius's canal from the Tiber to the sea, the canal from the Nile to the port of Alexandria, Odoacer's canal from Mentone, near Ravenna, to the sea, the Roman canals in England and Lombardy, the Moorish canals in Granada (which languished when Ferdinand conquered the country!) all indicate the knowledge the ancients had with this form of inland navigation.

The general early theory was to make inland navigation possible by means of a " canalization " of rivers. One of the most successful efforts in this direction is the Grand Canal of China, the great highway of the Middle Kingdom; it was built in the thirteenth century, to connect the waters of the Yang-ste and the Pei-ho Rivers, the former the great waterway of central China, the latter the waterway of the strategic province of Chili. This great work nearly a thousand miles in length is a series of canalized rivers. Other canals, such as that pushed forward by Charlemagne to unite the Rhine with the Danube, were almost impossible until the invention

of the lock. The blockheadedness of the
Spaniard is most clearly shown in the attitude of a certain state paper, though, in
fact, it very nearly voices a fundamental
scientific law; with reference to the canalization of Spanish rivers a decree of a state
council read: " .   . if it had pleased
God that these rivers should have been
navigable, he would not have wanted
human assistance to have made them such,
but that, as he has not done it, it is plain
that he did not think it proper that it should
be done. To attempt it, therefore, would
be to violate the decrees of his providence,
and to mend these imperfections which he
designedly left in his works." There was
a vast deal of mending the imperfections of
Providence before men found the secret of
one of Providence's simplest laws.

   In 1481 two engineers at Viterbo, Italy,
invented the canal lock by which craft
could be lifted or lowered from one level
to another. The discovery gave great im-
petus to canal building, especially in Italy.
The first canal in France was the Braire,
built in 1605–1642. The Orleans was
opened in 1675. Of all European works of

this character the Languedoc Canal, built by Riquet, from 1667 to 1881, was the most conspicuous. It connected the Bay of Biscay and the Mediterranean, its termini being Narbonne and Toulouse. It is one hundred and forty-eight miles in length and its summit level is six hundred feet above the sea, " while the works on its line embrace upwards of one hundred locks and fifty aqueducts, an undertaking which is a lasting monument of the skill and enterprise of its projectors; and with this work as a model it seems strange that Britain should not, till nearly a century after its execution, have been engaged in vigorously following so laudable an example." [1]

The Romans had built two canals in England, the Caer Dyke and the Foss Dyke; of the former only the name remains. " Camden in his *Britannia* states that the Foss Dyke was a cut originally made by the Romans, probably for water supply or drainage, and that it was deepened and rendered in some measure navigable in the year 1121 by Henry I. In 1762 it was reported on by Smeaton and

[1] *Encyclopedia Britannica*, " Canals."

Grundy, who found the depth at that time to be about two feet, eight inches. They, however, discouraged the idea of deepening by excavation. . . It was resolved [1840] to increase the dimensions of the canal, and to repair the whole work . . and thus that ancient canal, which is quoted by Telford and Nimmo as ' the oldest artificial canal in Britain,' was restored to a state of perfect efficiency, at a cost of £40,000.'' [2]

The internal navigation of Great Britain was the subject of legislation in 1423; locks were known on the river Lee as early as 1570. The seventeenth century saw considerable canal digging, but the island is so narrow that in early days the coasting trade and navigable rivers answered almost all purposes of commerce. About the middle of the eighteenth century the growth of manufacturing centers wrought great changes, and for half a century canal building in England came to the fore, though south of Durham no point was fifteen miles from navigation. The Duke of Bridgewater, procuring a grant for construction of canals

[2] *Id.*

in 1758, was the great promoter in this line
of industry at this period.   These were the
brilliant days of John Smeaton, civil engineer and improver of hydraulic machinery.
Born near Leeds in 1724, he achieved perhaps his most celebrated success in 1759, by
the completion of the Eddystone lighthouse.
His other famous works were building
Ramsgate Harbor and the Forth and Clyde
Canal in Scotland; this work, first proposed
by Charles II, was completed in 1789, according to Smeaton's plans.   It is thirty-five
miles in length, passing over a summit
level of one hundred and sixty feet, by
means of thirty-nine locks.   In Ireland the
Grand Canal from Dublin to Ballinasloe,
with a total length of one hundred and
sixty-four miles, was built in 1765.   In
1792 the Royal Canal leading from Dublin
to Tormansburg, ninety-two miles, was
completed.   Nearly five thousand miles of
canals have been built in Great Britain.

It was natural that an echo of the awaken-
ing of internal improvements in England
should have been heard in her American
colonies where such a vast field for such
enterprise lay awaiting a similar awaken-

ing.   It is believed that as early as 1750 a canal or sluice was dug in Orange County, New York, by Lieutenant-governor Colder for the transportation of stone.   The earliest planned lock canal in the provinces was the Schuylkill and Susquehanna, surveyed from the Schuylkill River near Reading, Pennsylvania, to Middletown on the Susquehanna in 1762.   Work on this canal was not begun until 1791, but only four miles were opened by 1794, when the work again paused.   Not until 1821 was it resumed, and the canal was completed in 1827 under the name of the Union Canal.   It became a division of the later Pennsylvania Canal.

The second canal survey in the American colonies was of a route between Chesapeake Bay and the Delaware River in 1764. A new survey was made of this proposed canal in 1769, under the auspices of the American Philosophical Society; it was not, however, until 1804 that work was commenced on this canal — the Chesapeake and Delaware, as it was known — and this was soon suspended.   The route was resurveyed in 1822 and completed, thirteen and one-half miles long, in 1829.

It is interesting to note that the subject of canals was being widely mooted in America at a time far remote from the day when they came actually into existence. England waited a century after the celebrated Languedoc Canal in France proved what vast good this form of internal improvement could bring, before she took up the canal problem in earnest. Within half a century, and less, after canal building was common in England it became common in young America. We were comparatively quick to make the most of opportunities in this as in every branch of invention and promotion which helped toward annihilating distances. The great extent of our territory in itself was an inducement to this end. Our colonial roads were often impassable in the winter season and wretched in any wet weather; the main line of communication was the Atlantic Coast, never easily navigated and, for a large part of the year, extremely dangerous in these early days before the invention of the blessings of our present coast surveys, lighthouses, and lightships. As a consequence, it was natural that the idea gained ground rapidly that if the splendid rivers which are scattered in profusion up and down our seaboard could be connected by canals a new era would dawn in our coast-wise trade, which was, in fact, almost our only trade.   Thus it came about that hosts of schemes were proposed for connecting our Atlantic rivers and bays.

In many cases our rivers were easily navigated for long distances into the interior; but these distances varied in different seasons of the year, and when, in the last quarter of the eighteenth century the western movement became prominent, and the rivers were ascended further than before, the question of the navigation of unnavigable waters came quickly to the fore.   Unfortunately for their pocket-books, our forefathers did not agree with the Spanish idea that improving unnavigable rivers was a wilful attempt " to mend the imperfections of providence."   The story of the sorry attempts to make such rivers as the Mohawk and upper Potomac navigable proves that the Spanish decree was somehow in the right, whether the Spanish reasoning was correct or not.

The following letter written by Benjamin Franklin to S. Rhoads, Mayor of Philadelphia, from London, August 22, 1772, with reference to the improvement of rivers and building of canals is an interesting early view of the subject. Mayor Rhoads had evidently applied for and received data respecting the canals of Great Britain:

" I think I before acknowledg'd your Favour of Feb. 29. I have since received that of May 30. I am glad my Canal Papers were agreeable to you. I fancy work of that kind is set on foot in America. I think it would be saving Money to engage by a handsome Salary an Engineer from here who has been accustomed to such Business. The many Canals on foot here under different great Masters, are daily raising a number of Pupils in the Art, some of whom may want Employment hereafter, and a single Mistake thro' Inexperience in such important Works, may cost much more than the Expense of Salary to an ingenious young Man already well acquainted with both Principles and Practice. This the Irish have learnt at a dear rate in the first Attempt of their great

Canal, and now are endeavoring to get
Smeaton to come and rectify their Errors.
With regard to your Question, whether it
is best to make the Schuylkill a part of the
Navigation   to   the   back   Country,   or
whether the difficulty of that River, subject to all the Inconveniences of Floods,
Ice, &c., will not be greater than the Expense of Digging, Locks, &c., I can only
say that here they look on the constant
Practicability of a Navigation, allowing
Boats to pass and repass at all Times and
Seasons, without Hindrance, to be a point
of the greatest Importance, and, therefore,
they seldom or ever use a River where it
can be avoided.   Locks in Rivers are subject to many more Accidents than those in
still water Canals; and the Carrying away
a few Locks by Freshets of Ice, not only
creates a great Expense, but interrupts
Business for a long time till repairs are
made, which may soon be destroyed again,
and thus the Carrying on a Course of Business by such a Navigation be discouraged,
as subject to frequent interruptions.   The
Toll, too, must be higher to pay for such
Repairs.   Rivers are ungovernable things,

especially in Hilly Countries. Canals are quiet and very manageable. Therefore they are often carried on here by the Sides of Rivers, only on ground above the Reach of Floods, no other Use being made of the Rivers than to supply occasionally the waste of water in the Canals. I warmly wish Success to every Attempt for Improvement of our dear Country.   .   ."

The Revolutionary War put an end to many plans for the improvement of Franklin's "dear country." Immediately after the close of the war, however, the various projects were again advanced here and there as the young republic began to grasp the great opportunities that lay before it. Among the most important early undertakings were those which looked forward to a new West and the need of lines of communication in advance of the rough roads which were the only avenues of commerce. The scheme of improving the rivers which rose in the Alleghenies, and connecting their heads with the waterways which flowed into the Ohio River at Lake Erie, was one of the moving projects of the hour. The improvement of the James, Potomac, and

Mohawk Rivers for this purpose com-
manded the attention of the nation at the
time; these projects were the first steps
toward building the Chesapeake and Ohio
and the Erie canals, and will be treated in
the chapters devoted to those topics.   It is
our purpose here only to emphasize in
general terms the mania for improving the
minor waterways in which so many millions of dollars were wasted before such
advice as that given by Franklin in 1772, as
quoted, was found to be well-grounded.

The spirit of this enterprising but unfortunate movement cannot be caught better
than by studying the papers and projects
of a " Society for promoting the improvement of roads and inland navigation,"
formed in Philadelphia at the beginning of
the last decade of the eighteenth century
and of which the able but unfortunate
Robert Morris was president.   Much of
Pennsylvania's leadership in works of improvement was due to the activity of this
organization.   One of the main objects of
the society is stated in a memorial to the
Pennsylvania Assembly dated February 7,
1791, the introduction of which reads:

" The memorial of ' The Society for promoting the improvement of roads and inland navigation,'

" *Refpectfully fheweth,*

" That your memoralifts, refiding in various parts of this ftate, with a view to contribute their beft endeavors to promote the internal trade, manufactures and population of their country, by facilitating every poffible communication between the different parts of the ftate, have lately formed themfelves into a fociety, by the name above mentioned. And knowing that the Legiflature, with the laudable intention of advancing the beft interefts of this commonwealth, and availing themfelves of the extenfive information, which they have obtained of the geography and fituation of the country, have now under their confideration the important fubject of roads and inland navigation; we, therefore, beg leave, with all poffible deference, to fuggeft fome important confiderations which have occurred to us in our enquiries into this fubject." A description of the position of Pennsylvania then follows, with an outline of her rivers which, as was then believed, were to become by improvement the commercial avenues of the dawning age. "To combine the interefts of all the parts of the ftate, and to cement them in a perpetual commercial and political union, by the improvement of thofe natural advantages, is one of the greateft works which can be fubmitted to *legiflative* wifdom; and the prefent moment is particularly aufpicious for the undertaking, and if neglected, the lofs will be hard to retrieve."[3]

Following this the river systems of Pennsylvania are taken up in order, showing the number of miles of waterways which it was supposed were capable of being connected and made avenues of trade. The two main divisions and their various subdivisions were as follows:

" *Delaware Navigation*

1.   From the tide water at Trenton falls to lake Otfego, the head of the northeaft branch of Sufquehanna

2.   From the tide water on Delaware to Ofwego on lake Ontario

[3]*An Historical Account of the Rise, Progress and Present State of the Canal Navigation in Pennsylvania* (Philadelphia, 1795), p. 1.

" *Sufquehanna Navigation*

1. From Philadelphia, or the tide waters of Schuylkill, to Pittfburgh on the Ohio

2. From Philadelphia to Prefqu'Ifle on lake Erie, by the Juniata and Kifkeminetas &c

3. From Philadelphia to Prefqu'Ifle, by the weft branch of Sufquehanna, Sinnemahoning and Conewango.

4. From Philadelphia to Prefqu'Ifle, by the weft branch of Sufquehanna, Sinnemahoning and Toby's creek.

5. From the tide waters of Sufquehanna to Pittfburgh.

6. From the tide waters of Potomack, at George Town, to Pittfburgh

7. From Conedeffago lake to New York

8. From the middle of the Geneffee country to New York "

The Pennsylvania Assembly responded liberally to the appeal of Robert Morris's society, and appropriated, April 13, 1791, £22,220 for the improvement of Pennsylvania rivers; the largest appropriations were for the " Sufquehanna, from Wright's ferry to the mouth of Swatara creek, inclu-

sive " £5,250; " For the river Delaware,"
£3,500; " For the river Schuylkill,"£2,500;
Conemaugh, £2,800; Allegheny, £150; and
Lehigh, £1000.[4]   Thus it will be seen that
the improvement of rivers was firmly considered to be one of the important undertakings of the day.

[4]*Id.*, pp. 73, 74.

CHAPTER II

THE POTOMAC COMPANY

GEORGE Washington's efforts to promote internal improvement in Virginia and Maryland with special reference to the Middle West have been lightly sketched in other portions of this work.[5] A more or less complete examination into the Potomac Company must be essayed here, for among the improvements of internal waterways in America that of the Potomac urged by Washington meant to the last quarter of the eighteenth century what the building of the Erie Canal meant to the first quarter of the nineteenth.

Having maintained with earnestness for many years that Virginia and Maryland should, through the Potomac River, secure the trade of the rising empire west of the

[5] *Historic Highways of America*, vols. iii, pp. 189-204; xii, pp. 15-30.

Alleghenies, Washington, at the close of the Revolution, gave himself wholly up to this commercial problem. Before peace was declared he left the Continental camp at Newberg and made a long, dangerous tour up the Mohawk Valley, examining carefully the portages to Wood Creek at Rome, and to Lake Otsego at Canajoharie. With that far-sighted shrewdness which, of itself, made him a marked man, he felt that this route which avoided the mountains was the great rival of his Potomac River. Yet he was no narrow partisan. Returning from his tour he wrote Chevalier de Chastellux from Princeton, October 12, 1783: " Prompted by these actual observations, I could not help taking a more extensive view of the vast inland navigation of these United States and could not but be struck with the immense extent and importance of it, and with the goodness of that Providence, which has dealt its favors to us with so profuse a hand. Would to God we may have wisdom enough to improve them. I shall not rest contented, till I have explored the western country, and traversed those lines, or great part of them, which have given bounds to a new empire.''

This clear cry of enthusiasm was from the heart, and within a year Washington carried out his plan of western exploration. Of this journey we had occasion to speak in our sketch of the Old Northwestern Turnpike.[6] In that connection our attention was confined to the portage route between the Cheat and Potomac Rivers. Here his plan for a water avenue from East to West must be emphasized as the first chapter in the history of both the Potomac Company and the Chesapeake and Ohio Canal. This cannot be so well done as by quoting the summary of the *Journal* of this trip, which has never been published.[7] It will be seen that it is the basis and, in part, the first draft of his famous Letter to Harrison written upon his return to Mount Vernon:[8]

''And tho' I was disappointed in one of the objects which induced me to undertake this journey namely to examine into the situation quality and advantages of the Land which I hold upon the Ohio and Great

[6]*Historic Highways of America*, vol. xii, ch. iii.
[7]*Id.*, note i.
[8] October, 1784.

Kanhawa — and to take measures for rescu-
ing them from the hands of Land Jobbers
& Speculators — who I had been informed
regardless of my legal & equitable rights,
Patents, &c^a; had enclosed them within
other Surveys & were offering them for
Sale at Philadelphia and in Europe.— I
say notwithstanding this disappointment I
am well pleased with my journey, as it has
been the means of my obtaining a knowledge of facts — coming at the temper &
disposition of the Western Inhabitants —
and making reflections thereon, which,
otherwise, must have been as wild, inco-
her^t, or perhaps as foreign from the truth,
as the inconsistency, of the reports which
I had received even from those to whom
most credit seemed due, generally were

" These reflections remain to be summed
up.

" The more then the Navigation of
Potomack is investigated, & duely considered, the greater the advantages arising
from them appear.—

" The South or principal branch of
Shannondoah at Mr Lewis's is, to traverse
the river, at least 150 Miles from its Mouth;

all of which, except the rapids between the Bloomery and Keys's ferry, now is, or very easily may be made navigable for inland Craft, and extended 30 Miles higher.—The South Branch of Potomack is already navigated from its Mouth to Fort Pleasant; which, as the Road goes, is 40 computed Miles; & the only difficulty in the way (and that a very trifling one) is just below the latter, where the River is hemmed in by the hills or mountains on each side — From hence, in the opinion of Col° Joseph Neville and others, it may, at the most trifling expense imaginable, be made navigable 50 Miles higher.—

" To say nothing then of the smaller Waters, such as Pattersons Creek, Cacapehen, Opeckon &c^a; which are more or less Navigable; — and of the branches on the Maryland side, these two alone (that is the South Branch & Shannondoah) would afford water transportation for all that fertile Country between the bleu ridge and the Alligany Mountains; which is immense — but how trifling when viewed upon that immeasurable scale, which is inviting our attention!

" The Ohio River embraces this Commonwealth from its Northern, almost to its Southern limits.— It is now, our western boundary.— & lyes nearly parallel to our exterior, & thickest settled Country.—

" Into this River French Creek, big bever Creek, Muskingham, Hockhocking, Scioto, and the two Miamas (in its upper Region) and many others (in the lower) pour themselves from the westward through one of the most fertile Country's of the Globe; by a long inland navigation; which, in its present state, is passable for Canoes and such other small craft as has, hitherto, been made use of for the Indian trade.—

" French Creek, down w[ch] I have myself come to Venango, from a lake near its source, is 15 Miles from Prisque Isle on lake Erie; and the Country betw[n] quite level.— Both big bever creek and Muskingham, communicates very nearly with Cuyahoga; which runs into lake Erie; the portage with the latter (I mean Muskingham) as appears by the Maps, is only one mile; and by many other acc[ts] very little further; and so level between, that the Indians and Traders, as is affirmed, always drag

their Canoes from one River to the
other when they go to War — to hunt, — or
trade. — The great Miame, which runs into
the Ohio, communicates with a River of the
same name, as also with Sandusky, which
empty themselves into lake Erie, by short
and easy Portages. — And all of these are
so many channels through which not only
the produce of the New States, contemplated by Congress, but the trade of *all* the
lakes, quite to that of the Wood, may be
conducted according to my information, and
judgment — at least by one of the Routs —
thro' a shorter, easier, and less expensive
communication than either of those which
are now, or have been used with Canada,
New Y^k or New Orleans. —

" That this may not appear an assertion,
or even an opinion unsupported, I will
examine matters impartially, and endeavour
to state facts. —

" Detroit is a point, thro' which the
Trade of the Lakes Huron, & all those
above it, must pass, if it centres in any
State of the Union; or goes to Canada;
unless it should pass by the River Outa-
wais, which disgorges itself into the S^t

Lawrence at Montreal and which necessity only can compel; as it is from all acc^ts longer and of more difficult navigation than the St Lawrence itself.—

" To do this, the Waters which empty into the Ohio on the East Side, & which communicate nearest and best with those which run into the Atlantic, must also be delineated —

" These are, Monongahela and its branches, viz, Yohiogany & Cheat.— and the little and great Kanhawas; and Greenbrier which emptys into the latter.—

" The first (unfortunately for us) is within the jurisdiction of Pensylvania from its Mouth to the fork of the Cheat, indeed 2 Miles higher — as (which is more to be regretted) the Yohiogany also is, till it crosses the line of Maryland; these Rivers I am persuaded, afford *much* the shortest Routs from the Lakes to the tide water of the Atlantic, but one not under our controul; being subject to a power whose interest is opposed to the extension of their navigation, as it would be the inevitable means of withdrawing from Philadelphia all the trade of that part of its

western territory, which, lyes beyond the
Laurel hill.— Though any attempt of that
Government to restrain it I am equally
well persuaded w^d cause a separation of
their territory; there being sensible men
among them who have it in contemplation
at this moment.— but this by the by.— The
little Kanhawa, which stands next in order,
& by Hutchins's table of distances (between
Fort Pit and the Mouth of the River Ohio)
is 184½ Miles below the Monongahela, is
navigable between 40 and 50 Miles up, to
a place called Bullstown.— Thence there
is a Portage of 9½ Miles to the West
fork of Monongahela — Thence along the
same to the Mouth of Cheat River, and up
it to the Dunker bottom; from whence a
portage may be had to the N^o branch of
Potomack.

" Next to the little, is the great Kanhawa; which by the above Table is 98½
miles still lower down the Ohio.— This is
a fine Navigable river to the Falls; the
practicability of opening which, seems to
be little understood; but most assuredly
ought to be investigated.

" These then are the ways by which the

Produce of that Country; & the peltry and
fur trade of the Lakes may be introduced
into this State; & into Maryl^d; which
stands upon similar ground.— There are
ways, more difficult & expensive indeed by
which they can also be carried to Philadelphia — all of which, with the Rout to Albany, & Montreal,— and the distances by
Land, and Water, from place to place, as far
as can be ascertained by the best Maps now
extant — by actual Surveys made since the
publication of them — and the information
of intelligent persons — will appear as follow — from Detroit — which is a point, as
has been observed, as unfavourable for us
to compute from (being upon the North
Western extremity of the United territory)
as any beyond Lake Erie can be.—

viz —

From Detroit to Alexandria

is

To Cuyahoga River . . .	125	Miles
Up the same to the Portage .	60	
Portage to Bever C^k . .	8	
Down Bever C^k to the Ohio	85	
Up the Ohio to Fort Pitt .	25	303

The Mouth of Yohiogany . 15
Falls to Ditto   .   .   . 50
     Portage   .   .   . 1
Three forks or Turkey foot . 8
Fort Cumberl[d] or Wills Creek 30
Alexandria 200   304

     Total   607

To Fort Pitt — as above   .   .   303
The Mouth of Cheat River . 75
Up it, to the Dunker bottom 25
North branch of Potomack . 20
Fort Cumberland   .   . 40
Alexandria 200   360

To Alexand[a] by this Rout   .   663

From Detroit to Alexandria avoiding
Pensylvania *

To the M[o] of Cuyahoga   .   . 125 Miles
The carrying place with ⎫
  Muskingham River ⎬ 54
Portage   .   .   . 1
The M[o] of Muskingham   . 192
The little Kanhawa   .   . 12   384

* The Mouth of Cheat River & 2 Miles up it is in
Pensyl[a]

Up the same . . .	40	
Portage to the West Bra	10	
Down Monongahela to Cheat	80	
Up Cheat to the Dunker Bot^m	25	
Portage to the N° bra.		
Potom^k . . .	20	
Fort Cumberland . .	40	
Alexandria	200	415

Total by this Rout . . 799

From Detroit to Richmond

To the Mouth of the little Kan-		
hawa as above . . .		384
The Great Kanhawa by Hut-		
chins's Table of Distances	98½	
The Falls of the Kanhawa		
from information . .	90	
A portage (supp^e) .	10	
The Mouth of Green brier & up		
it to the Portage . .	50	
Portage to James R^r .	33	281

Richmond 175

Total . . . 840

Note — This Rout *may be* more incorrect

than either of the foregoing, as I had only
the Maps, and vague information for the
Portages — and for the distances from the
Mouth of the Kanhawa to the Carrying
place with Jacksons (that is James) River
and the length of that River from the Carrying place to Richmond — the length of
the carrying place above is also taken from
the Map tho' from Information one would
have called it not more than 20 Miles.

From Detroit to Philadelphia
is

		Miles
To Presque Isle . . .		245
Portage to Lebeauf . .	15	
Down french Creek to Venango	75	
Along the Ohio to Toby's Creek	25	115
To the head spring of D⁰ . .	45	
By a Strait line to the nearest		
Water of Susqueᵃ . .	15	
Down the same to the West		
branch	50	
Fort Augusta at the Fork . .	125	
Mackees (or Mackoneys) Cᵏ .	12	
Up this	25	
By a strait line to Schuylkˡ .	15	

Reading	.	.	.	.	32	
Philadelphia	.	.	.	.	62	381

Total	.	.	.	741

By another Rout

To Fort Pitt as before	.	.	.	303
Up the Ohio to Tobys C^k	.	.		95
Thence to Phil^a as above	.	.		381

Total	.	.	.	.	779

Note — The distances of places from the
Mouth of Tobys Creek to Philad^a are taken
wholly from a comparative view of Evan's
and Sculls Maps — The number, and length
of the Portages, are not attempted to be
given with greater exactness than these —
and for want of more competent Knowledge, they are taken by a strait line between
the sources of the different Waters which
by the Maps have the nearest communication with each other — consequently, these
Routs, if there is any truth in the Maps,
must be longer than the given distances —
particularly in the Portages, or Land
part of the Transportation, because no
Road among Mount^{ns} can be strait — or

waters navigable to their fountain heads.
From Detroit to Albany
is

To Fort Erie, at the N° end of Lake Erie	350	
Fort Niagara — 18 Miles of wch is Land transpⁿ . .	30	380
Oswego		175
Fall of Onondaga River .	12	
Portage . . .	1	
Oneida Lake by Water . .	40	
Length of D° to Wood C^k .	18	
Wood C^k very small and Crooked	25	
Portage to Mohawk .	1	97
Down it to the Portage .	60	
Portage . . .	1	
Schenectady	55	
Portage to Albany .	15	131
In all		783
To the City of New York .		160
Total		943

From Detroit to Montreal
is

To Fort Niagara as above . .		380
North end of Lake Ontario	. 225	
Oswegatche . . .	. 60	
Montreal — very rapid .	. 110	395
In all		775
To Quebec		180
Total		955

" Admitting the preceding Statement, which as has been observed is given from the best and most authentic Maps and papers in my possession — from information — and partly from observation, to be tolerably just, it would be nugatory to go about to prove that the Country within, and bordering upon the Lakes Erie, Huron, & Michigan would be more convenient when they came to be settled — or that they would embrace with avidity our Markets, if we should remove the obstructions which are at present in the way to them.—

" It may be said, because it has been said, & because there are some examples of it in proof, that the Country of Kentucke, about the Falls, and even much higher up the Ohio, have carried flour and other articles to New Orleans — but from whence has it proceeded? — Will any one who has ever calculated the difference between Water & Land transportation wonder at this? — especially in an infant settlement where the people are poor and weak handed — and pay more regard to their ease than to loss of time, or any other circumstance?

" Hitherto, the people of the Western Country having had no excitements to Industry, labour very little; — the luxuriancy of the Soil, with very little culture, produces provisions in abundance — these supplies the wants of the encreasing population — and the Spaniards when pressed by want have given high prices for flour — other articles they reject; & at times, (contrary I think to sound policy) shut their ports against them altogether — but let us open a good communication with the Settlements west of us — extend the inland Navigation as far as it can be done with convenience — and shew them by this means, how easy it is to bring the produce of their Lands to our Markets, and see how astonishingly our exports will be encreased; and these States benefitted in a commercial point of view — wch alone is an object of such Magnitude as to claim our closest attention — but when the subject is considered in a political point of view, it appears of much greater importance.''

By means of letters, urging these private speculations on public attention, to Governor Harrison and James Madison, the matter of improvement of the Potomac was brought before the Virginia legislature. The consent and coöperation of Maryland being of greatest importance, General Washington, General Gates, and Colonel Blackburn were appointed by the legislature to obtain the concurrent action of the Maryland legislature. On December 20, 1784, the deputation, with the exception of Colonel Blackburn who was detained by illness, reached the Maryland capital. A committee from that state being duly appointed to confer upon the matter in hand, a conclusion was reached as contained in the following report.

'' That it is the opinion of this conference, that the removing the obstructions in the River Potomac, and the making the same capable of navigation from tide-water as far up the north branch of the said river as may be convenient and practicable, will increase the commerce of the commonwealth of Virginia and State of Maryland, and greatly promote the political interests of the United States, by forming a free and easy communication and connection with the people settled on the western waters, already very considerable in their numbers, and rapidly increasing, from the mildness of the climate and the fertility of the soil.

"That it is the opinion of the conference, that the proposal to establish a company for opening the River Potomac, merits the approbation of, and deserves to be patronized by, Virginia and Maryland; and that a similar law ought to be passed by the legislatures of the two governments to promote and encourage so laudable an undertaking." [9]   It was further agreed that the commonwealths of Virginia and Mary-

[9] Pickell's *A New Chapter in the Early Life of Washington*, p. 44.

land should each subscribe for fifty shares of stock in the undertaking in order to "encourage individuals to embark in the measure" and as "a substantial proof to our brethren of the western territory of our disposition to connect ourselves with them by the strongest bonds of friendship and mutual interest." How closely Washington's plan was carried out is suggested in the following resolutions: "That it is the opinion of this conference, from the best information they have obtained, that a road, to begin about the mouth of Stony River, may be carried in about twenty or twenty-two miles to the Dunker Bottom or Cheat River; from whence this conference are of opinion, that batteaux navigation may be made, though, perhaps, at considerable expense. That if such navigation cannot be effected by continuing the road about twenty miles further, it would intersect the Monongahela where the navigation is good, and has long been practiced. . . That it is a general opinion, that the navigation in the Potomac may be extended to the most convenient point below, or even above the mouth of Stony River, from whence to set-off a road to Cheat River; and this conference is satisfied that that road, from the nature of the country through which it may pass, wholly through Virginia and Maryland, will be much better than a road can be made at any reasonable expense from Fort Cumberland to the Youghiogheny, which must be carried through Pennsylvania." In a succeeding resolution it is affirmed that the Dunkard Bottom route is more feasible than one from Fort Cumberland to Turkey Foot [Connellsville, Pennsylvania], though the latter road, if improved, would be of great value to many settlers upon and near it. The legislatures of the respective states were asked to appoint examiners to view the doubtful portions of the South Branch (from Cumberland to the mouth of Stony Creek) and the Cheat (from established navigation and Dunkard Bottom) and lay out a road between the heads of practicable navigation on each. It was also suggested that Virginia and Maryland ask permission of Pennsylvania to lay out a road from Cumberland to Turkey Foot on the Youghiogheny.

Accordingly Virginia and Maryland passed laws authorizing the formation of a company for the improvement of the Potomac River. " I have now the pleasure," wrote Washington to Richard Henry Lee, February 8, 1785, " to inform you that the Assemblies of Virginia and Maryland have enacted laws, of which the inclosed is a copy.[10]   They are exactly similar in both States.   At the same time, and at the joint and equal expense of the two governments, the sum of six thousand six hundred and sixty-six dollars and two thirds is voted for opening and keeping in repair a road from the highest practicable navigation of this river to that of the River Cheat, or Monongahela, as commissioners, who are appointed to survey and lay out the same, shall find most convenient and beneficial to the western settlers." Washington believed fully that the project was to be a great success for stockholders; he estimated that they would receive twenty per cent from investments in Potomac improvement in a few years.[11]

[10] See appendix A, p. 219.

[11] E. Watson's *History of the . . Western Canals in the State of New York*, p. 87.

The subscription books of the new company having been, as the law required, opened on February 8, 1785, a summons was issued for a meeting of the subscribers at Alexandria, Virginia, on May 17. The meeting having been called to order, Daniel Carrol was elected chairman and Charles Lee, clerk.[12] The books being opened, it was found that Virginia had subscribed for two hundred and sixty-six shares, the Richmond book showing one hundred shares, the Alexandria book, one hundred and thirty-five, and the Winchester book thirty-one; Maryland had subscribed for one hundred and thirty-seven shares, divided as follows: Annapolis, seventy - three; Georgetown, forty - two; Frederick, twenty-two. The total shares were therefore four hundred and three, giving the company a capital of £40,300. President and four directors of the Potomac Company, as it was known, being ballotted for, George Washington was elected president, and Thomas Johnson, Thomas Sim

[12] All particulars concerning the inner history of the Potomac Company are from Pickell's *A New Chapter in the Early Life of Washington*; the author had access to all documents in the case.

Lee, John Fitzgerald, and George Gilpin were elected directors.

The services of Mr. James Rumsey, the mechanician, being secured, as general manager of improvements, the president, directors and manager made an examination of the river with a view to planning the work to be done. Three important impediments to navigation were immediately attacked; these were known as " Great Falls," " Seneca Falls " and " Shenandoah Falls." The " Great Falls " of these early days are the rapids and falls above Washington which bear the same name today. Seneca Falls were early known as " Sinegar Falls," in the Revolutionary era on Fry and Jefferson's map. They lie just above Great Falls, near the mouth of Seneca Creek. Shenandoah Falls were at the present Harper's Ferry at the mouth of the river of the same name. In the summer of 1785 parties of workmen were blasting and removing the boulders at these two points until the fall rains put an end to the work. Attention was then given to excavating a canal around Great Falls, concerning which there was a great diversity of opinion, especially as to its lower termination.

The work of this first season quickly brought out the fact that it was a great task which the company had undertaken. This may have been the reason why payment on shares had been so slow; already the company's treasury was almost depleted. " The original motive which actuated the stockholders seemed for some cause to have abated, and it required the *master spirit* of the enterprise to be exerted, to prevent at this important and critical juncture, a total abandonment of the project. . . The State of Maryland had failed to pay the sums due on the shares it held, and a large number of individual stockholders had also neglected to meet their instalments. . . The treasury was no longer able to liquidate the claims of individuals against it, and a total prostration of its credit seemed inevitable unless soon relieved." [18]

Strenuous efforts on the part of the officers brought desirable results and with

[18]*A New Chapter in the Early Life of Washington*, pp. 83–84.

the opening of the season of 1786 work on the improvements was being pushed with earnestness. At the annual meeting, August 7, the same officers were re-elected and the treasurer's books were examined and found in good order. The president and directors were allowed thirty shillings in Virginia currency for the time they had spent in the business of the company. It was determined that the directors should visit in person the river from Great Falls upward, to inspect the ground, choose a channel, and take such action as, in their judgment, the case demanded. This was done, the trip covering four days; and as a result the legislatures of the states were requested to extend the time limit from three years to November 17, 1790. In this the legislatures acquiesced.[14]

During the new fiscal year additional difficulties arose to unite with those of unpropitious weather and financial distress to delay and discourage. The appointment of Mr. Richardson Stewart as " principal Assistant Superintendent " resulted in the resignation of Mr. Rumsey, who

[14] Hening's *Statutes at Large*, vol. xii, ch. cxiv.

preferred charges of incompetency against Mr. Stewart.[15] The directors replied to the charges in the order they were made, finding Mr. Stewart guilty of only one, namely of "causing another servant to burn Michael Barnet with a hot iron without reason;" the directors declared, without fear or favor, "that in this Mr. Stewart acted with an impropriety the Board disapproves of"![16] A difficulty had arisen, early in the work, in securing workmen and in keeping them in submission to law and order when once obtained. In the fall of 1785 half the laborers were dismissed from the company's service. The secretary of the company now, and at numerous times thereafter, was in correspondence with parties in Baltimore (Messrs. Stewart and Plunket) and in Philadelphia (Mr. John Maxwell Nesbit) who might secure workmen for the Potomac improvements.[17] Furnishing the workmen with liquors also seems to have been a troublesome item. In November, 1785, a contract was made

[15] *A New Chapter in the Early Life of Washington*, pp. 94–95.
[16] *Id.*, p. 98.
[17] *Id.*, p. 78.

with William Lyles and Company to furnish
" what rum might be necessary for the use
of the hands employed by the company on
the river " at the rate of two shillings per
gallon.[18]   For the winter 1786–7 the manager was directed to retain such a force as
was deemed necessary at a monthly wage
(from November 12 to April 12) of thirty-
two shillings for common laborers, and
forty shillings for " prime hands, with the
usual ration except spirits, and with such
reasonable allowance of spirits as the
manager may from time to time think
proper.   .   ." [19]

At a meeting of the directors January 3,
1787, the financial crisis was faced sternly.
The funds were quite exhausted and work
would have to be suspended unless the
delinquent stockholders immediately advanced the assessment long overdue.   It
was determined to warn delinquents that
unless advances were made within the next
five months the legal recourse of reselling
subscribed stock at auction would be

[18]*Id.*, p. 83.   Cf. *Historic Highways of America,*
vol. v, p. 142.

[19]Pickell, *ut supra*, p. 100.

resorted to. A few responded but the
" large majority continued delinquent." In
accordance with the threat of the directors,
it was announced in public advertisements
that forty-six shares of stock in the Potomac Company would be offered at auction
at the court house at Alexandria on Mon-
day, May 14, and nine shares at Shuter's
tavern in Georgetown on May 21. The
attitude of the general public toward the
Potomac improvement scheme was revealed
clearly at these auctions — for at neither
Alexandria nor Georgetown was a single
bid made when these shares were offered
for sale, though numbers of people had
gathered out of interest or curiosity.[20]

A meeting of the board of directors was
called at the mouth of the Shenandoah
(Harper's Ferry) June 2, 1788, at which it
was determined to cut down expenses
" without jeopardizing the progress of the
work." It was now the opinion of the
board that by the ensuing season loaded
boats could descend from the pool or
" reach " above Seneca Falls to tide-water;
this meant that a channel in Seneca Falls

[20]*Id.*, p. 104.

had been opened and the canal about Great Falls completed. It was given out that in July the entire force of workmen would be concentrated at Shenandoah Falls to hasten the opening of a channel at that point. At the annual meeting in Alexandria, August 4, it was reported that high water had delayed operations but that by November 1, the channel would be open from tidewater to Cumberland. Since the last meeting of the company £2,990 sterling had been paid into the treasury, making a total of paid up assessments to date of £13,719 18s 8d.

The election of a president for the year ensuing was postponed, as it was plain that Washington was soon to be named President of something of more note than a Potomac Company. In May 1787 he had been elected president of the National Convention at Philadelphia, and it was clear that he would be first choice as executive of the new republic. He was elected President of the United States for the term beginning March 4, 1789. From the day of his withdrawal from the Potomac Company its affairs languished — proving clearly that but for Washington's name and energy the organization would probably never have existed.

On ten different occasions did the legislatures of Virginia and Maryland extend the time demanded by law for the completion of the Potomac improvements, between 1786 and 1820. By this time the promotion of the Erie Canal aroused the proprietors to inquire into the feasibility of cutting a canal from the Potomac to the Ohio River. During thirty-six years $729,380 had been spent in the attempt to improve the Potomac and little had been accomplished; an inquiry into the affairs of the Potomac Company by a state commission, appointed in 1821, and reporting July, 1822, resulted in the following report: " . . that the affairs of the Potomac Company have failed to comply with the terms and conditions of the charter; that there was no reasonable ground to expect that they would be able to effect the objects of their incorporation; that they have not only expended their capital stock and the tolls received, with the exception of a small dividend of five dollars and fifty cents on each share declared in 1802, but had incurred a heavy debt which their resources would never enable them to discharge; that the floods and freshets nevertheless gave the only navigation that was enjoyed; that the whole time when produce and goods could be stream bourne on the Potomac in the course of an entire year, did not exceed forty-five days; that it would be imprudent and inexpedient to give further aid to the Potomac Company." The committee advised a more effectual method of inland navigation and suggested the plan of a canal through the region in which the Potomac Company had proposed to operate, to be connected with Baltimore, the metropolis of the Chesapeake, by means of a lateral canal, from some point along the Potomac Valley.

CHAPTER III

IT is exceedingly interesting to note that the old plan of Washington's, by which the Middle West and Northwest were to be held in fee by those who controlled the Potomac, was as dominant now in 1823 as it was, within a limited circle, in 1784. In fact this is what the Potomac Company, the Potomac Canal Company, the Chesapeake and Ohio Canal Company and the Baltimore and Ohio Railway Company have all stood for — this commercial control of the trans-Allegheny empire. Our general plan demands a full examination of this phase of our subject at the time at which we now have arrived — 1823.

In view of the canal project now on the tapis the Potomac Company adopted a resolution on February 3, 1823, signifying their willingness (!) to surrender their charter on liberal terms to a new company for the prosecution of the new plan of communication. A bill was introduced, in accordance with the same plan in the Maryland legislature, to incorporate a joint stock company to be known as " The Potomac Canal Company." It was estimated that the proposed work of cutting a canal, from Potomac tide-water (Washington, D. C.) up the valley, across the mountains to a branch of the Ohio, and down the same, at a million and a half dollars, of which Virginia, Maryland, and the District of Columbia were each to subscribe one-third."[21]

A commission was appointed by Virginia and Maryland to examine the old route across the Alleghenies marked out by Washington with a view to the possibility of constructing a canal from the head of the Potomac to one of the heads of the Ohio. James Schriver made an examination of the Alleghenies with reference to the new canal in the summer of 1823, and the result was given to the public in the form of a report entitled: *An Account of Surveys and Examinations with Remarks and Documents*

[21] Scharf's, *History of Maryland*, vol. iii, p. 156.

relative to the projected Chesapeake and Ohio, and Ohio and Lake Erie Canals.[22]   Though we find Mr. Schriver a United States Associate Civil Engineer in 1826,[23] he seems to have made his explorations "to satisfy himself and a few friends."[24]   Since the day of Washington's explorations in 1784 it was generally understood that the most practicable route for a road or canal from the Potomac to an Ohio tributary would follow the portage route outlined by Washington from the Potomac at the mouth of Savage River to the Cheat River.   But the emphasis given by Washington to this portage was not based wholly on utilitarian motives.   He desired his route to keep within the bounds of Virginia and Maryland — the possessors of the Potomac — for any more northerly course would carry the route into Pennsylvania.   Washington, however, was searching for waterways which could be made navigable; Schriver, a generation later, sought only for streams which could furnish sufficient water for a

[22] Baltimore, 1824.

[23] *House Docs. no. 10, 19th Cong., 2d. Sess.*, p. 9.

[24] *An Account of Surveys and Examinations*, p. 3.

canal. As a result, Schriver was satisfied with the head of the Youghiogheny which, though it could never be made navigable, yet contained plenty of water to fill a canal. Schriver's proposed route, therefore, left the Potomac at the mouth of Savage River, ascended that stream and its tributary, Crabtree Creek. Reaching Hinch's Spring by means of a tunnel,[25] the canal would follow the North Fork of Deep Creek and Deep Creek itself to the Youghiogheny. Descending the Youghiogheny and Monongahela, the Ohio River would be reached at Pittsburg. The vital question was thought to be whether there was a sufficient current of water to supply the summit level of the canal at the tunnel under Little Backbone Mountain.

The bill to incorporate the Potomac Canal Company, however, failed to pass the

[25] The probable success of a tunnel of a mile and a half in length was not doubted at this time. The Trent-Mersey Canal in England had five tunnels in ninety-three miles, and one (at Harecastle) was more than a mile and a half long, and over two hundred feet beneath the surface of the earth. Its cost was £31 10s 8d per yard. The Chesterfield canal had a tunnel at Hartshill three thousand yards long.—*An Account of Surveys and Examinations*, p. 57, *note*.

Maryland legislature. This brings us at
once face to face with one of the most
interesting phases of the subject — the
position and commanding influence of Baltimore in the commercial world at that
day. "The progress of the [Potomac
Canal Company] bill," writes the Maryland
historian Scharf, "caused much excite-
ment in Baltimore. The people of that
city, notwithstanding they were in favor of
internal improvements, and had freely subscribed for the construction of roads,
bridges, etc., were unanimously opposed to
this bill, because it called for an appropriation of the funds or credit of the State (one-
third of which they would be compelled to
pay) to an object that would be rather an
injury than a benefit to the trade of the
city. Though they had but a fortieth part
of the power of legislation in the House of
Delegates, they paid one-third part of the
taxes of the State, and as the funds of the
State were not sufficient to meet the ordinary expenses of about $30,000 a year, the
financial burden bore with great pressure
upon them. Besides, they especially objected to the Potomac canal, because, under

the bill in question, the canal was to termi-
nate as at present, in Georgetown, and the
privilege was virtually denied them of tap-
ping it so as to connect it by a canal with
Baltimore, if they so desired; besides, the
State was asked to cede to the company all
its rights to the waters of the river [Poto-
mac], thus virtually preventing the future
connection of the canal with the City of
Baltimore.   To produce concert of action
in the next session of the Maryland and
Virginia legislatures, the friends of the
measure began to hold meetings in various
parts of the country.   .   . [These
meetings] resulted in the assembling of a
convention in the city of Washington, on
Thursday, the 6th day of November, 1823,
with delegates from Maryland, Virginia,
Pennsylvania, and District of Columbia."[26]

The business of the convention, of which
Congressman Joseph Kent was chosen
chairman, was to advocate the enlargement
of the plan of the Potomac Canal Company
so that it would include Baltimore as its
eastern terminus, by means of a lateral
canal or an extension of the main canal

[26]*History of Maryland*, vol. iii, pp. 156–157.

from its terminus at Georgetown.
"Whereas, a connection of the Atlantic
and Western waters, by a canal," read the
introduction to the resolutions adopted,
"leading from the seat of the general
government to the river Ohio, regarded as
a local object, is one of the highest importance to the states immediately interested
therein, and, considered in a national
view, is of inestimable consequence to the
future union, security, and happiness of
the United States:

"*Resolved, unanimously*, That it is expedient to substitute, for the present defec-
tive navigation of the Potomac river above
tide water, a navigable canal, by Cumberland
to the mouth of Savage Creek, at the eastern
base of the Alleghany, and to extend
such canal, as soon thereafter as practicable, to the highest constant steamboat
navigation of the Monongahela or Ohio
river."[27]   Another resolution outlined a
plan of enlargement of the Potomac Canal
Company by the appointment of committees "each consisting of five delegates, to
prepare and present, in behalf of this

[27]*Niles Register*, vol. xxv, p. 173.

assembly, and in co-operation with the central committee, hereinafter provided, suitable memorials to the congress of the United States, and the legislatures of the several states before named [Virginia, Maryland, Pennsylvania, and the District of Columbia], requesting their concurrence in the incorporation of such a company and their co-operation, if necessary, in the subscription of funds for the completion of the said canal: And whereas, by an act of the general assembly of Virginia, which passed the 22d February, 1823, entitled, ' an act incorporating the Potomac canal company,' the assent of that state, so far as the limits of her territory render it necessary, is already given to this *object*, and for *its* enlargement to the extent required by the preceding resolution, the said act appears to furnish, with proper amendments, a sufficient basis: *Be it, therefore, resolved* That it will be expedient to accept the same as a charter for the proposed company, with the following modifications, viz: That in reference to its enlarged purpose, the name be changed to the ' Chesapeake and Ohio Canal.' '' These resolutions[28] are practically embodied in the act incorporating the Chesapeake and Ohio Canal.[29]

The two hundred delegates concluded their convocation by a banquet at Brown's Hotel, Washington, on Saturday evening. Certain of the " spontaneous sentiments " were: By the Secretary of State, John Quincy Adams, "the first right and the first duty of nations — self-dependence and self-improvement;" by the Secretary of War John C. Calhoun, " Canal navigation between the Atlantic and the western waters, essentially connected with the commerce, the defence, and the union of the states — may it receive the patronage and support of the nation;" by C. F. Mercer, soon to be the first president of the Chesapeake and Ohio Canal, "the eastern and western country — whom the Author of Nature has joined together, may no man put asunder;" by Mr. James Schriver, pioneer surveyor on the upper Potomac, " The Chesapeake and Ohio; they have ' passed meeting' [30]—

[28] *Id.*, pp. 173–175.

[29] See note 33.

[30] " Passed meeting," a practice among the Friends previous to the marriage ceremony.

may their marriage be speedily consummated.'' A toast which tells of Clay's presidential ambitions was proposed by B. S. Forrest of Maryland after the speaker's withdrawal from the board in the following technical phrase: '' Henry Clay, qualified to pass the summit level; neither giddy in ascending, nor dismayed in descending!'' The members of the important Central Committee were Charles F. Mercer, John Mason, Walter Jones, Thomas Swann, John McLean, William H. Fitzhugh, H. L. Opie, Alfred H. Powell, P. C. Pendleton, A. Fenwick, John Lee, Frisby Tilghman, and Robert W. Bowie. The committee to memorialize Congress was as follows: Walter Jones, John Mason, George Washington Park Custis, Robert I. Taylor, S. H. Smith.[31]

That George Washington's original plan of connecting the Potomac with the Great Lakes was still dominant, a resolution of this convention proves; the Virginians and Marylanders were bound to control the commerce of the Lakes even with the Erie

[31]*National Intelligencer; Niles Register*, vol. xxv, p. 175.

Canal as a rival. Their resolution read:
"*And be it further resolved*, That a committee of five delegates be appointed to prepare, and cause to be presented, in behalf of this convention, a suitable memorial to the state of Ohio, soliciting the co-operation of that state in the completion of the Chesapeake and Ohio canal, and its ultimate connexion with the navigation of Lake Erie; and that, for the latter purpose, the memorial shall respectfully suggest the expediency of causing the country, between the northernmost bend of the river Ohio, and the southern shore of Lake Erie, together with the waters of Great Beaver and Cayuga [Cuyahoga] creeks, and all other intervening waters near the said route, to be carefully surveyed, with the view of ascertaining the practicability and probable cost of a canal, which, fed by the latter, shall connect the former."[32] Mr. Schriver, in his volume quoted, gives much attention to this western extension of the Chesapeake and Ohio Canal. "The proposed *Ohio and Lake Erie Canal*," he affirms, " is intimately blended

[32] *Id.*, pp. 174–175.

with that of the *Chesapeake and Ohio*. In
the opinion of many, it is embraced and
constitutes only a part of the same grand
design; but whether it be considered in
connexion with it, or independently, it is
confessedly a project of vast public importance, involving considerations of great
national and local concern."

The Washington canal convention
brought forth much fruit; its demands
were eminently reasonable; the plan of
operations proposed was logical, and fair
to all concerned. The Potomac Canal
Company could not face the future successfully without the friendship of Maryland and
Maryland's commercial metropolis. The
legislature of Virginia passed an act incorporating the Chesapeake and Ohio Canal
Company, January 27, 1824.[33] Upon being
slightly amended, it was passed by the
Maryland legislature January 31, 1825. A
perusal of the act will show that the new
company was capitalized at $6,000,000,
divided into 60,000 shares of $100 each.
Certificates of stock in the old Potomac
Company, or debts of the same, were to be

[33] See appendix B, p. 225.

accepted at par or nominal value for certificates in the new company, under certain conditions and limitations. The canal was divided into Eastern and Western sections, the mouth of the Savage River being the division point;[34] if the company did not begin work in two years, or if one hundred miles were not completed in full in five years, the charter should become null and void. If the western section was not begun within two years after the time allowed for the completion of the eastern section, or was not completed in six years, the right and title of the company " in said western section, shall cease and determine." It will be noted that failure to complete the western section did not affect the company's right to the eastern section. The annual dividends were not to exceed fifteen per cent, and unless one-fourth of the capital should be subscribed all subscriptions were to be void.

In December 1823 President Monroe

[34] While the law divided the canal into only two sections, eastern and western, the engineers divided it into three, eastern, middle, and western. The two former met at Cumberland, and the latter began at the mouth of Casselman's River.

presented the internal improvement proposed by the Chesapeake and Ohio Canal
Company to Congress, and in April 1824
an appropriation of $30,000 was made to
procure surveys and estimates in order to
prove the feasibility of the plan.   In May
the President appointed Brigadier-general
Simon Bernard and Lieutenant-colonel
Totten and Civil Engineer John L. Sullivan
of Massachusetts as a board to outline the
most suitable route for a canal from Potomac tide-water to the Ohio River.   Their
report was made October 23, 1826.[35]   The
four memoirs of the report include a survey
of the Potomac Valley from tide-water to
Cumberland, Maryland, by Lieutenant-
colonel J. J. Abert; a descriptive statement with reference to the eastern section
of the summit level between the Potomac
and the heads of the Ohio by Captain
William G. McNeill; a descriptive account of Casselman's River or the Somerset route, also by Captain McNeill; a
review of other routes by James Schriver.
   In the eastern section the canal was
planned on the Maryland side of the

[35] *State Papers 19th Cong., 2d Sess., Doc. no. 10.*

Potomac River, because the obstacles on that bank were of less magnitude than those on the opposite Virginia shore, the exposure was more favorable, an earlier navigation could be secured there in the spring, and a later navigation in the fall, and no aqueduct would be required at Cumberland, as Wills Creek enters the Potomac at that point from the Maryland shore. Moreover the water supply from Maryland to the Potomac exceeded that of Virginia, the rivers of the latter sending 190 cubic feet of water per second into the Potomac, and the former 267.35 cubic feet. While perhaps not fully accurate, these figures approximated the truth.

The length of the eastern section was placed at one hundred and eighty-six miles, and it was divided into eleven subdivisions marked by the following points beginning at Cumberland: South Branch of the Potomac, Great Cacapan, Licking Creek, Great Conococheague, Antietam Creek, one mile below Harper's Ferry, Monocacy River, Seneca Creek, Great Falls, Little Falls, and Georgetown. The old canal of the Potomac Company was to be used by

the new canal as far as possible. A summary of the eastern section reads:

Distance (miles) 185⅝
Descent (feet) 578
Number of locks   74
Estimated cost   .   .   . $8,177,081.05

In seeking a route across the towering
ridges between the Potomac and heads of
the Ohio, the course first suggested by
Washington and studied by commissioners
since his day was discarded by the board of
surveyors which now planned the actual
course of the canal. The Chesapeake and
Ohio Canal Company being incorporated in
Pennsylvania, it was now no object to
keep the highway within the territories of
Virginia and Maryland alone. Upon
exploration, it was found that a route up
Wills Creek from the Potomac at Cumberland, Maryland, and across to Casselman's
River, a branch of the Youghiogheny, was
a more favorable route than that by way of
Savage River and Deep Creek to the
Youghiogheny. The question was determined by the supply of water at summit
level. The reservoirs in the Deep Creek
plan would have to be twelve miles in

length, while those by the more northerly route would be but three and one-half miles in length. A saving of six million cubic yards of water by evaporation in the Casselman's route made that way far more advantageous. The lockage on the Deep Creek route was eight hundred and seventy-three feet more than by the Casselman route; on the other hand this was equalized by the fact that the tunnel on the latter route was to be four miles and eighty yards long, while the Dewickman tunnel on the Deep Creek route was only one mile and five hundred and sixty-eight yards long. With all factors taken into account, it was estimated that the Deep Creek route would cost $2,861,288.90, and the Casselman's or Flaugherty Creek route $2,324,315.37, or more than a half million dollars less than the Deep Creek route.

This Middle Section, therefore, extended from Cumberland, or the western extremity of the Eastern Section, to the mouth of Casselman's River in the Youghiogheny, the "Turkey Foot" of pioneer days.[36] Its length was seventy miles and one thousand

[34] See *Historic Highways of America*, vol. iii, p. 133.

and ten yards. The lockage was nineteen hundred and sixty-one feet and the summit was to be crossed by a tunnel four miles and eighty yards long, dug at eight hundred and fifty-six feet below the summit of the ridge. The Middle Section was divided into an eastern and a western portion. The former had two subdivisions; the first, descending from the summit, was fifteen miles in length, with a descent of one thousand and sixteen feet, from the eastern end of the summit level to the mouth of Little Wills Creek; the second subdivision, nearly fourteen miles long, and with a descent three hundred and nine feet, extended from Little Wills Creek to the western end of the Eastern Section, below Cumberland. The western portion of the Middle Section was, likewise, divided into two subdivisions; the first, sixteen miles long with a drop of two hundred and sixteen feet, ran from the western end of the summit level to the mouth of Middle Fork Creek; the second, nineteen miles long, with a descent of four hundred and twenty feet, ran from there to the mouth of Casselman's. The summit level was

five miles and one thousand two hundred
and eighty yards long, to be crossed by a
tunnel four miles and eighty yards long
and a deep cut the remaining distance.

The summary of the Middle Section
reads:

	Distance		Ascent and descent	Number of locks	Estimated cost
	miles	yds.			
Eastern Portion	29	240	1325	166	$3,856,623.60
Summit Level	5	1280			3,471,967 01
Western Portion	35	1250	636	80	2,699,532.25
	70	1010	1961	246	10,028,122.86

The Western Section began four hundred
and forty yards below the junction of Casselman's River with the Youghiogheny and
extended to Pittsburg on the Ohio River.
The canal was planned on the right bank
of the Youghiogheny and Monongahela
Rivers, and was divided into four subdivisions:

Termini	Miles	Descent (feet)
Western end of Middle Section to Con- nellsville	27½	432
Connellsville to Sewickly Creek . .	27¼	144
Sewickly Creek to mouth of Youghiogheny	16½	8
Mouth of Youghiogheny to Pittsburg .	14	35

The summary for the Western Section was:

Distance (miles)	85¼
Descent (feet)	. 619
Number of locks . . .	. 78
Estimated cost . .	. $4,170,223.78

The total estimate of the board, therefore, for the entire work was as follows:

	Distance miles.	Ascent and descent yds.	Number of locks	Estimated cost
Eastern Section .	185	1078	578	74	$8,177,081.05
Middle Section .	70	1010	1961	246	10,028.122.86
Western Section .	85	348	619	78	4,170,223.78
Totals . .	341	676	3158	398	$22,375,427.69 [37]

Under the head of " General Considerations " [38] the board treated minutely the proposition presented by the acts incorporating the Chesapeake and Ohio Canal Company, and the treatise is one of the most interesting studies of early commerce between the East and the West. The great population and area concerned on both sides of the Alleghenies, the increased value of real estate which would follow the

[37] *Id.*, p. 62.
[38] *Id.*, pp. 65–80.

building of the canal, the articles of import and export which would pass and repass over the great highway, the probable revenue which would be derived from tolls, the enhanced value, commercially, of a canal to the Ohio River whenever the Ohio was in turn connected with Lake Erie, and the strategic military position and value of the canal on the shortest route from Atlantic tide-water to the Ohio River and the Great Lakes by way of the national capital, are points considered at some length.

This report of the board, naming over twenty millions as the cost of the canal was an overwhelming and disappointing surprise. The capital of the Chesapeake and Ohio Canal Company was, as we have seen, only six million — in itself a tremendous sum in that day. The blow fell heavily on Baltimore; while the building of the canal in the Potomac Valley was entirely reasonable, it was the larger interests of the great scheme that had a special appeal to the capitalists of the Maryland metropolis. As a highway between tide-water and the Ohio Basin the scheme had been greatly favored by them. Already, on March 6, 1825, the Maryland legislature had provided for the formation of what was known as the " Maryland Canal Company " with a capital of half a million dollars, which should bind the Chesapeake and Ohio Canal with the city of Baltimore. In any lesser sense — as merely a canal in the Potomac Valley — the Chesapeake and Ohio Canal was of far less interest to Baltimoreans than an improvement of communications, for instance, to the rich Susquehanna country. And the moment it was known that merely the Middle Section, of seventy miles, of the Chesapeake and Ohio Canal was to cost nearly twice the entire proposed capitalization of the company, the idea of a continental canal to the West through the Alleghenies was deemed impracticable at Baltimore. A new estimate of the expense was undertaken by James Geddes and Nathan S. Roberts, who cut down the figures named by General Bernard one half.[39] Those greatly interested in the advancement of the scheme hailed

[39] Scharf's *History of Maryland*, vol. iii, p. 165.

this announcement with delight, but the more conservative relied upon the estimates of the national board as being the most reliable.

The actual resulting effects of the discouraging report of the board concerning the cost of this enterprise were so far-reaching, that it is altogether proper to pause a moment here and consider the position and influence of the city of Baltimore, and note what the failure of the canal scheme meant to her. As a commercial metropolis Baltimore's reputation was very great, and second only to that of Philadelphia. Not only was it a great seaboard market, but throughout the preceding half century it had been one of the great markets for western produce. Its position was unique; although a seaport it was many miles nearer the Ohio Valley than any rival. In laying out possible landward routes from the Ohio River to the seaboard for the Cumberland National Road, the commissioners found that the route to Baltimore was thirty-nine miles shorter than to Philadelphia, and forty-two miles shorter than to Richmond. The distance from the seaport of Baltimore to Brownsville, Pennsylvania, on the Monongahela, where navigation by boat was almost always possible, was only two hundred and eighteen miles. Thus Baltimore was the natural eastern metropolis for the trade of the West. Moreover, Baltimore had, up to date, taken perhaps all advantages of her situation, and had grown rich in consequence; the building of the Cumberland Road had been of great benefit, for Cumberland was but the half-way house to Baltimore. Baltimore and Maryland had improved their opportunities by building many miles of fine roads, really extending the Cumberland Road to Baltimore and tide-water.

Baltimore's commercial prestige was secure so far as land ways were concerned. New measures calling for water ways now on foot, made popular by the great success of the Erie Canal, promised to overturn all previous considerations. The coach and freighter, it seemed, were now to be replaced by the easy-gliding canal-boat. Baltimore had been the metropolis for western trade during the reign of the freighter. Must she resign her place upon

the advent of the canal-boat? This was
the question which was being agitated
throughout the years of the Potomac
chimera; the failure of that scheme again
restored the confidence of the Baltimoreans.
But the revival of the plan under the new
arrangement of a canal from tide-water to
the Ohio Basin again created alarm. The
position of the Marylanders in this extremity is well indicated by one of Niles's
editorials as follows: " The ' National Intelligencer' of Tuesday last [November,
1823], in an article signed ' Multum in
Parvo,' contains a very illiberal attack on
the people of Baltimore, because of their
supposed opposition to the Potomac canal.
It accuses us of ' avarice and ambition ' —
of being ' selfish ' — as ' jealous ' of Washington, and as preparing to oppose a resto-
ration of their ' political rights' to the
people of the District of Columbia! It also
puts it down as *impossible* to conduct an arm
of this canal to our city. . . *Baltimore* ' avaricious and ambitious! ' We refer
to the support afforded by loans, and the
great disbursements made on our own re-
sponsibility, during the late war; the splen-

did public roads with which we have inter-
sected the country, and the beautiful edi-
fices, fountains, &c. that we have built in
our city, in proof of our ' avarice ' — and
direct public attention to North Point and
Fort McHenry, for evidences of our ' ambition:' and, as to being ' selfish ' or
' jealous,' these are nearly the last things
that should be said about Baltimore;   .   .
So far as my information goes,   .   .   the
citizens of Baltimore are not *opposed* to the
Potomac canal: but how is it possible to
expect their *support* for it when the following facts are considered:

" 1.   We have expended a million of
dollars on certain public roads, to obtain
that trade which the canal is designed to
deprive us of.

" 2.   Yet, and notwithstanding we are
to suffer this loss of capital and trade, if
the canal should be made as heretofore
proposed, we must pay *one third* of Maryland's share of the expense of making it:
that is to say, 10,000 dollars a year will *be
added* to the amount of our taxes, though
such is our present condition that the usual
taxes can hardly be collected, through the

depreciation of property and want of business.

" 3.   .   .   As well might we accuse the people of the District of Columbia of *selfishness*, because they will not help us to make a canal to the Susquehannah, as they can censure us for preferring that canal to one on the Potomac.   We are willing that the Potomac canal should be made — but not at our cost; until, at least, we have fully ascertained what can be done in respect to a favorite measure of the same nature.   But we must be permitted to doubt whether the people of the district would feel very zealous about the navigation of the Potomac, provided it was ascertained as practicable, and *conditioned*, that an arm of the canal should be extended to Baltimore, though the last is so much nearer the sea than Washington, &c." [40]

As noted, Maryland refused to pass the bill incorporating the Potomac Canal Company, because of the objections, largely, of Baltimoreans.   To the enlarged plan embraced under the name of Chesapeake and Ohio Canal Company, assent was given,

[40] *Niles Register*, vol. xxv, p. 145.

under the impression that full connection with the West by canal was possible, and that Baltimore was to become, virtually, the eastern terminus. The report of the national board as to the enormous expense of the canal precluded the thought of the building of the Middle and Western Sections, and, consequently, deprived it of its genuinely national character. The discouragements discovered by the Maryland Canal Company in their attempt to find a satisfactory location for a canal route from the Potomac to Baltimore,[41] also had its effect in strengthening the opinion of Baltimore capitalists that Baltimore could never hold the trade of the West by water routes as for half a century she had held it by land routes. New York and Philadelphia were fast surpassing her, and, by means of the Pennsylvania and Erie Canals, seemed in a fair way to secure the trade of the West which once had been hers. In the editorial already quoted the discouraging state of trade in Baltimore is hinted at.

Philip E. Thomas, president of the Mechanic's Bank of Baltimore and a commis-

[41] Scharf's *History of Maryland*, vol. iii, p. 164.

sioner for Maryland for the Chesapeake and
Ohio Canal Company, resigned his office
upon reviewing the report of General
Barnard, and, calling into his counsels
George Brown, the two in private faced the
situation in which Baltimore was placed.
Without hope of taking any advantage of
the Potomac to gain the trade with the
West, with New York and Pennsylvania
fast outstripping Baltimore in trade and
population and both pushing canals to the
West, the outlook for Baltimore seemed
unpromising indeed.   These two energetic
and   daring   men, in comparatively a
moment's time, changed the whole com-
plexion of affairs, and brought not only the
eyes of the world to Baltimore but in very
fact brought back to her the commercial
prestige, so far as western trade was concerned, which she had enjoyed in the day
of the stagecoach and freighter.   On the
twelfth of February, 1827, the plans of
Thomas and Brown had gone so far that a
meeting at the home of Mr. Thomas of
over a score of Baltimore merchants and
promoters was called " to take into consideration the best means of restoring to

the City of Baltimore that portion of the Western Trade which has lately been diverted from it by the introduction of steam navigation and other causes."

The plan of Thomas and Brown comprehended the building of a railway from Baltimore to the Ohio. Both men had brothers in England who had forwarded reports of railway experiments there. The matter had received considerable previous attention and the great proposition was discussed with an intense interest. From all the data which were gathered by the correspondents abroad, the proposition was wholly reasonable. And in its realization the promoters would find a relish intensified a hundred-fold, because of the rumors circulated that Baltimore must resign her commercial position to Alexandria or Georgetown because of the building and influence of the Chesapeake and Ohio Canal. The two railways then in operation in the United States were at Quincy, Massachusetts, a road to a quarry; and at Mauch Chunk, Pennsylvania, from the Lehigh River to the Summit Coal Mine, nine miles distant. As means of conveying heavy freight rapidly the success of the "rail road" was assured. The idea of conveying passengers was an afterthought; it was the freight traffic that Baltimore had lost — it was the freight traffic which the Chesapeake and Ohio Canal would draw from even the best roads Baltimore could build or have built. By means of rails, cars with freight could be moved, it was estimated, at least twelve miles an hour, and railroads could be built anywhere macadamized roads could go. The supply of water at the summit level was not a critical factor.

The result of this meeting at the home of Mr. Thomas was the appointment of a committee which was ordered to review the whole proposition, and report a plan of action. On February 19, the committee report was ready, and the second meeting was held. The report affirmed that rail roads promised to "supercede Canals as effectually as Canals have superceded Turn- pike Roads," and recommended that "a double Rail Road" be constructed "be- tween the City of Baltimore and some suitable point upon the Ohio River, by the most eligible and direct route, and that a Charter to incorporate a Company to execute this work be obtained as early as practicable." [42]   On February 28, 1827, a charter was granted by the Maryland legislature; it was confirmed by Virginia on March 8, and by Pennsylvania February 22, 1828. Mr. Thomas resigned the presidency of the Mechanic's Bank to give his whole attention to the affairs of the enterprise.

A unique situation now presents itself to the historical inquirer. On the one hand we find the Chesapeake and Ohio Canal Company, under the presidency of Charles F. Mercer of Virginia, chairman of the Committee of Roads and Canals of the National House of Representatives, backed by a capital of over three and one-half million dollars, ready to proceed in building a canal through the Potomac Valley from Washington to Cumberland; on the other hand is the new rail road company called the Baltimore and Ohio Rail Road Company, with Mr. Thomas at its head, backed, in 1828, by four millions of dol-

[43] W. P. Smith's *A History and Description of the Baltimore and Ohio Rail Road* (Baltimore, 1853), p. 13.

lars, beginning to build a rail road from Baltimore to the Potomac Valley, up the valley to Cumberland, and across the mountains to the Ohio River. It was evident at the start that the rivalry would be tremendously bitter; that the two companies would give rise to factions which would harm and decry each other in every way possible. The canal idea was, comparatively, very new, and the Erie Canal being successfully prosecuted from the Hudson to the Lakes had created immense enthusiasm. On the other hand the rail road was almost an untried novelty; on such roads as were in operation in England and America horse power was the only power to be relied upon; sails were in use but were not successful under many circumstances. The steam engine had not been successfully adapted as yet; the roadbeds were far more costly than even the most expensive macadamized roads; there was still a question whether the mountains could be spanned by this method of transportation, and whether, even if the locomotive could be utilized on a straight track, it could ever be useful on a curved track!

The bitterness of the rivalry was intensified by the fact that the two companies were organized within the same states, to operate in exactly the same territory and both seeking the same carrying trade. And, lastly, one company had its origin in a detrimental report from the highest authority made concerning the other. The seed of the Baltimore and Ohio Railway lay in General Bernard's report of 1826, in which the cost of building the Chesapeake and Ohio Canal across the mountains was estimated at a prohibitive figure.

Both companies went to work eagerly, and both sure of success. The infancy of the rail road science, and the fact that as yet nothing had been done in all the world on such a scale as was proposed by the Baltimore and Ohio Rail Road, naturally rendered public opinion more or less skeptical; while as for the canal, success was practically assured. It would be taking a very narrow outlook upon the situation to describe the building of the canal, without presenting a briefly sketched-in history of its great rival for western trade. The two must go hand in hand.

Early in 1828, both companies were in the field surveying the route of their two highways. At the point of conflict, where the railway approached the Potomac River, it was easily seen that trouble would be precipitated. In fact, as early as June 10, the canal company got from Judge Buchanan an injunction against the railway company, to prevent them from encroaching upon lands needed by the former and granted them by charter rights.[43] The railway company returned the compliment by obtaining an injunction from the " chancellor of the state of Maryland " likewise restraining the canal company.[44] " . . If we understand it [the situation]," wrote the perplexed editor of the *Register*, June 28, " the state of things is as it was, before the injunction obtained of Judge Buchanan."[45] The canal promoters' view of the affair was thus voiced by the editor of the *National Journal*: " It appears to us to be very essential to the harmonious prosecution of these two great works, that

[43] *Niles Register*, vol. xxxiv, p. 266.
[44] *Id.*, p. 282.
[45] *Id.*

the rights of each company should be precisely defined. It was with this view, we believe, that the injunction [of June 10] in the present stage was applied for; in order that the question how far the charter granted to the canal company, giving to them the privileges of condemning such land as may be necessary for the construction of that work, barred any other company from obtaining land along the same line, until the objects of the canal company should be accomplished. By the final settlement of this question, in the beginning, all ground for future collision would be removed. We should regret, therefore, if our Baltimore neighbors should regard as an act of hostility to them, that which is, in fact, simply an assertion of our own rights. There is no disposition to embarrass their work, to which we desire all success; there is no wish to delay it, as is evident from the offer which is said to have been made . . to refer the question to . . the court of appeals now sitting at Annapolis.'' [46]

Plans were already making by the rival

[46] *Id.*, p. 267.

companies for grand celebrations on the Fourth of July succeeding, when, near Washington, the ground should be broken for the canal, and, at Baltimore, the " Foundation of the Rail Road," in the shape of the corner-stone, should be laid. These rival celebrations attracted great crowds to the two cities on the day named. At Washington the streets were alive with people at an early hour, and at seven o'clock the directors of the Chesapeake and Ohio Canal Company met the honored guests of the day at Tilley's Hotel. These included the President of the United States and cabinet, and the various ambassadors of foreign countries then in the city, and other dignitaries, including survivors of the Revolutionary War. The procession, attended by troops and regaled with the music of bands, marched to the Potomac and embarked on the steamboat " Surprize," for a journey to the Great Falls of the Potomac. Crowds followed on either bank of the river. " The sun shone now and then from the clear blue heavens through the fleecy clouds," wrote the inspired reporter of the *National Intelli-*

gencer; all nature " seemed to smile upon
the scene." Disembarking, the company
marched to canal-boats lying in the old
canal built by the indefatigable labors of
Washington's Potomac Company nearly fifty
years before. During the journey up the
canal, we are assured, the " senses of the
company were regaled by a scene at once
novel and really enchanting. . . There
was a part of this passage, when the music
of Moore's sweet song of ' The meeting of
the waters,' poured its melody on the ear
so as to suspend the labor of the boatmen,
and charm to silence every voice." Two
companies of riflemen saluted the arrival
of President Adams on the ground. " Thousands hung upon the overlooking hill to
the north, and many climbed the umbrag-
eous trees." Within a hollow square, surrounded by the crowds, a spot was marked
for the raising of the first spadeful of earth
by John Quincy Adams. Then " amidst a
silence so intense as to chasten the anima-
tion of hope and to hallow the enthusiasm
of joy," the mayor of Georgetown handed
Mr. Mercer, president of the Chesapeake
and Ohio Canal Company, the implement

with which the ground should be broken.

" There are moments," said Mr. Mercer,
" in the progress of time, which are the
counters of whole ages. There are events,
the monuments of which, surviving every
other memorial of human existence, eternize
the nation to whose history they belong,
after all other vestiges of its glory have
disappeared from the globe. At such a
moment have we now arrived. Such a
monument we are now to found." At this
point Mr. Mercer handed the spade to
President Adams who, in turn, delivered
the address of the day. In the course of
his oration the speaker said: " To subdue
the earth is pre-eminently the purpose of
the undertaking, to the accomplishment of
which the first stroke of the spade is now
to be struck. That it is to be struck by
this hand, I invite you to witness." At
this point the President attempted to sink
the spade into the ground; but it struck a
root. " Not deterred by trifling obstacles," wrote an eye-witness, " from doing
what he had deliberately resolved to perform, Mr. Adams tried it again, with no
better success. Thus foiled, he threw

down the spade, hastily stripped off and laid aside his coat, and went seriously *to work*. . . The multitude . . raised a loud and unanimous cheering, which continued for sometime after Mr. Adams had mastered the difficulty." [47]

Simultaneously with this memorable celebration, an imposing ceremony was being enacted at Baltimore. " Fortunately," we read in the *Baltimore American*, " the morning of the fourth rose not only bright but cool, to the great comfort of the immense throng of spectators that, from a very early hour, filled every window in Baltimore street, and the pavement below, from beyond Bond street on the east, far west on Baltimore street extended, a distance of about two miles." It was estimated that seventy thousand people were in attendance. During the early morning the crowds streamed toward the spot about two miles from the city, just south of the Frederick turnpike, where on a rise of ground in the open field a pavilion was raised for the reception of the honored guests of the occasion. The distance of

[47]*Id.*, vol. xxxiv, pp. 325-328.

the scene of laying the corner-stone of the Baltimore and Ohio Rail Road from Baltimore made the processional display more imposing, led by the First Baltimore Hussars.

The venerable guest of the day was Charles Carroll of Carrollton, the only surviving signer of the Declaration of Independence. After the invocation and the reading of the Declaration of Independence, John B. Morris, a director of the Baltimore and Ohio Rail Road, addressed the assembled throng. His words were singularly prophetic. "We are about opening the channel," he said, "through which the commerce of the mighty country beyond the Alleghany [Mountains] must seek the ocean — we are about affording facilities of intercourse between the east and west, which will bind the one more closely to the other, beyond the power of an increased population or sectional differences to disunite. We are in fact commencing a new era in our history; . . It is but a few years since the introduction of steam boats effected powerful changes, and made those neighbors, who were before far distant from each other. Of a similar and equally important effect will be the Baltimore and Ohio rail road. While the one will have stemmed the torrent of the Mississippi, the other will have surmounted and reduced the heights of the Alleghany.   .   . It is not in mortals to command success, but if a determination to yield to no obstacle which human exertion can overcome .   . can ensure success — success shall be ours."

Then, descending from his seat in the pavilion, Charles Carroll lifted a spadeful of earth from the designated resting place of the foundation stone, which was then set in position. Within the stone was placed a copy of the charter of the company, the newspapers of the day, and a scroll containing these words:

" This stone is deposited in commemoration of the commencement of the Baltimore and Ohio Railroad, a work of deep and vital interest to the American people. Its accomplishment will confer the most important benefits upon this nation, by facilitating its commerce, diffusing and extending its social intercourse, and perpetuating the happy union of these confederated states. The first general meeting of the citizens of Baltimore to confer upon the adoption of proper measures for undertaking this magnificent work, was on the 2d day of February, 1827. An act of incorporation, by the state of Maryland, was granted February 28th, 1827, and was confirmed by the state of Virginia March 8th, 1827. Stock was subscribed, to provide funds for its execution, April 1st, 1827. The first board of directors was elected April 23, 1827. The company was organized, 24th April, 1827. An examination of the country was commenced under the direction of lieutenant colonel Stephen H. Long and captain William G. McNeill, United States' topographical engineers, and William Howard, United States' civil engineer, assisted by lieutenants Barney, Trimble and Dillahunty of the U. S. artillery, and Mr. Harrison, July 2d. 1827. The actual surveys to determine the route, were begun by the same officers, with the additional assistance of lieutenants Cook, Gwynn, Hazzard, Fessenden, and Thompson and Mr. Guion, November 20th, 1827. The charter of the company was confirmed by the state of Pennsylvania, February 22d, 1828. The state of Maryland became a stockholder in the company, by subscribing for half a million of dollars of its stock March 6th, 1828. And the construction of the road was commenced July 4th, 1828, under the management of the following named board of directors: Philip Evan Thomas, president, Charles Carroll of Carrollton, William Patterson, Robert Oliver, Alexander Brown, Isaac M'Kim, William Lorman, George Hoffman, John B. Morris, Talbot Jones, William Steuwart, Solomon Etting, Patrick Macauley, George Brown, treasurer. The engineers and assistant engineers in the service of the company are, Philip Evan Thomas, president, Lieutenant-colonel Stephen Harryman Long, Jonathan Knight, Board of Engineers. Captain William Gibbs McNeill, U. S. topographical engineer. Lieutenants William Cook, Joshua Barney, Walter Gwynn, Isaac Trimble, Richard Edward Hazzard, John N. Dillahunty of the U. S. artillery. Casper Willis Weaver, superintendent of construction." [48]

[48] *Id.*, pp. 317–318.

Both companies now went quickly to work on their undertakings; in the same issue of the *Register* (July 19, 1828), and side by side on the same page are these notices:

" The engineers of the Baltimore and Ohio Rail Road Company have, by public notice, invited proposals for the construction of *twelve miles* of the road, commencing at the city [Baltimore] line, and extending westwardly." [49]

" The Chesapeake and Ohio Canal Company have issued proposals for the excavation, embankment and walling, of the 11½ miles of the Chesapeake and Ohio Canal, in half mile sections, extending from the head of the Little Falls to the head of the Great Falls of the Potomac river."

In August, thirty-four sections of the canal from Little Falls to Seneca (seventeen miles) were placed under contract and on September 1, work was actually begun.[50] " At this time the capital stock subscribed and payable in current funds, exclusive of

[49]*Id.*, p. 331.

[50]*Report of the Chesapeake and Ohio Canal Company* (for 1851), pp. 1-44.

subscriptions in the stocks and debts of the Potomac [Canal] Company, amounted to 36,094 shares, or $3,609,400 as follows:

	Shares	Equivalent to
United States .	. 10,000	$1,000,000
Washington City	. 10,000	1,000,000
Maryland .	. 5,000	500,000
Alexandria .	. 2,500	250,000
Georgetown .	. 2,500	250,000
Shephardstown .	20	2,000
Individuals .	. 6,074	607,400
	36,094	$3,609,400 " [51]

Though the rail road was far more of an experiment than the canal, its stock had been taken up quickly. " The subscription books of the company," reads a note in the *Register* of April 7, 1827, " were closed on Saturday the 31st ult. on which day alone were taken thirteen thousand three hundred and eighty-seven shares, making, with those previously taken, *forty-one thousand seven hundred and eighty-eight shares*, inclusive of the five thousand allotted to and taken by the corporation of Baltimore. The amount of money, therefore, sub-

[51] Scharf's *History of Maryland*, vol. iii, p. 170.

scribed by this city [Baltimore] alone, is *four millions one hundred and seventy-eight thousand dollars*, divided amongst *twenty-two thousand names*. . . Each name will be entitled but to 7-10ths of a share . . which will be further reduced by the subscriptions in Frederick and Hagerstown, which are not yet ascertained, but are supposed to amount to two thousand shares. It is believed that of this subscription, which outruns so largely the fund contemplated to be raised, but a comparatively small part has been made with a view to speculation. There is, therefore, every reason to think, that the stock is principally in the hands of persons who intend and are able to hold it." [52]

The question of stock subscription brings up one of the points of conflict between the canal and the road — a government subscription to the Baltimore and Ohio Rail Road. We have seen that the government had subscribed for ten thousand shares, or one million dollars, in the Chesapeake and Ohio Canal Company stocks. Accordingly, the Board of Directors, headed by Charles

[52] *Id.*, vol. xxxii p. 100 (from the Baltimore *American*).

Carroll, signed a memorial, January, 1828, to the United States Congress asking for a national subscription. " The Senate committee to which the memorial was referred reported a bill authorizing a subscription of $1,000,000. The committee of the House of Representatives also made a favorable report, but it being late in the session when the committee reported, it would submit no bill. The company therefore renewed its petition at the next session of Congress in 1829, but, although the committees of both houses of Congress recommended a qualified subscription to the company, the measure failed. It was said at the time[53] that the reason the company was unsuccessful in this application was because of the opposition of the president of the Chesapeake and Ohio Canal Company, who was at this time chairman of the committee on roads and canals in the House of Representatives.[54]

[53] Smith's *History and Description of the Baltimore and Ohio Rail Road*, p. 22.

[54] Reizenstein's, " The Economic History of the Baltimore and Ohio Railroad," *Johns Hopkins University Studies*, fifteenth series, vii–viii, p. 23; *Congressional Debates*, vol. vi (1829–30), pp. 453–455, 1136–1137.

The rail road company was not in great need of a national subscription, though dark days were at hand. A perusal of the reports of President Thomas, the first of which was made October 1, 1827,[55] will cause the reader to marvel "that the formidable obstacles almost daily encountered . . did not crush the energies of the Company, and induce them to abandon the work. . ." An unforeseen difficulty in the shape of an immense cut near Baltimore called for an expenditure of nearly a quarter of a million. And it soon developed that the Canal Company, which had deprived the rail road of the government's aid was yet to strike a harder blow.

By its charter the Chesapeake and Ohio Canal Company had secured a right of way for a canal on the Maryland bank of the Potomac from Washington to Cumberland. By its surveys the rail road was compelled to gain the Potomac at the "Point of Rocks," twelve miles below Harper's Ferry, and follow the river to that point. Otherwise a tunnel would have to be built under the mountain spurs — a financially

[55]*Niles Register*, vol. xxxiii, pp. 137-138.

impossible alternative.   The point at issue
in the great quarrel, which became exceedingly bitter and was at last settled only by
Federal interference, was, therefore, very
plain.   This famous dispute for right of
way through these strategic twelve miles
was not settled until 1832, both companies
suffering in consequence of the delay, and
the railway losing its argument but effect-
ing a compromise.   In this year the Court
of Appeals reversed the decision of the
Chancery Court of Maryland and sustained
the Canal Company's contention for the
right of way between the Point of Rocks
and Harper's Ferry.   After a series of
compromise proposals by the rail road to
the canal had been refused, the Maryland
legislature took up the matter, both works
being important to that commonwealth.
On May 9, 1833 a compromise was effected
by the passage of a law calling for the joint
construction of canal and rail road through
the disputed territory; to Messrs. Charles
F. Mayer and Bene S. Pigman great credit
was due in handling successfully this problem, which had at its root the bitter
rivalry of many years standing.   The com-

THE CACTOCIN AQUEDUCT

[By this aqueduct the Chesapeake and Ohio Canal crosses Cactocin Creek, ten miles from Harper's Ferry. It was over the right of way here at the "Point of Rocks" that the bitter quarrel between the canal and the Baltimore and Ohio Railway was precipitated. The piers of the railway bridge over the same stream may be seen through the first arch of the aqueduct]

promise cost the rail road heavily. It was
to subscribe for 2,500 shares of Canal Company stock ($266,000) and the canal com-
pany built the road through the territory
in dispute (the Point of Rocks). The Rail
Road Company completed the road to the
Maryland shore of the Potomac opposite
Harper's Ferry in 1834, it being opened
December 1. Here, however, it was to
pause, for the compromise signed by the
two companies demanded that the rail road
should not be built up the Potomac until
the canal should have been completed to
Cumberland — if that was done within the
time named in the charter (1840).

Though at all times master of the situation, the Canal Company found its task
tremendously heavy; the weather, varying
prices of labor and necessaries, combined
with great physical obstacles, rendered the
undertaking one in which patience was as
necessary as capital. Both were many
times exhausted. We have seen that contracts were first let in 1828. By the president's report to the legislature in January,
1831, we find that forty-eight miles were

under contract and that twenty-one miles were in use during the fall of 1830 and winter of 1830–31.[56] In February, 1833 the state of Virginia authorized a subscription of Chesapeake and Ohio Canal stock to the amount of $250,000, subject only to reasonable conditions.[57] In March, 1834, Maryland authorized an additional subscription of $125,000, and promised a larger subscription in case the National Government voted the investment of an additional million in the canal. Neither the government or any state, save only Maryland, befriended the Chesapeake and Ohio Canal, however, from this date forward.[58] At this time (June, 1834) the canal had cost $4,062,991.25. Seventy-eight miles remained to be built and the Company's funds were unequal to the task. The friends of the great work met in conven-

[56]*Report from the President of the Chesapeake and Ohio Canal Company to the Legislature of Maryland, January 31, 1831.*

[57]*Report to the Stockholders .   .* made *February 27th, 1851,* p. 47. Many of the following facts are taken from this *Report,* which is the only history of the Chesapeake and Ohio Canal Company extant. It will be referred to as *Report of 1851.*

[58]*Id.,* p. 48.

tion at Baltimore in the December following, and a committee was appointed to
report on probable expense of completion
of the canal, and committees to memorialize
Congress and the legislatures of states
interested in the work. The former committee brought in a report, which consisted
of nothing more reliable than an expres-
sion of opinion based on former experi-
ences, which gave the public to understand
that the canal could be completed in two
years with two million dollars.[59] The other
states turning a deaf ear to the plea, Maryland came to the rescue, March 7, 1835,
and appropriated the entire two millions
needed.[60] It was granted in the form of a
loan, the state reserving the power to con-
vert it into capital stock at any future time
if it was deemed expedient.

The company now took the steps which
should have preceded the circulation of any
opinion by friends of the canal as to the
expense of completing it — a survey and
estimate was made. This being done, it
was found that the cost of completing the

[59] *Id.*, p. 50.
[60] *Id.*, p. 52.

canal exceeded three and one-half millions. The consternation aroused by this report can be imagined. Many felt that Maryland had been deceived and imposed upon. But the friends of internal improvements arose to the occasion. Meetings were held up and down the state. The canal and rail road people united hands which formerly had been clinched in threatening attitude, and on June 3, 1835, the Maryland Legislature passed the famous "Eight Million Dollar Bill." [61] Its items were as follows:

To the Chesapeake and Ohio Canal Company . . .	$3,000,000
To the Baltimore and Ohio Rail Road Company . . .	3,000,000
To the Eastern Shore Rail Road Company	1,000,000
To the Maryland Canal Company	500,000
To the Annapolis Canal Company	500,000
	$8,000,000

As it stood the bill was a great victory for the Baltimore and Ohio Rail Road interests, as one of its most important pro-

[61] *Id.*, p. 61 (*Laws of 1835*, ch. 395).

visions demanded that the Chesapeake and Ohio Canal Company permit its rival to ascend the Potomac Valley.[62] Baltimore went wild over the passage of the act. A public dinner, fireworks, the ringing of bells, and a salute of a hundred guns gave evidence of the feeling at the Maryland metropolis. "The citizens of Baltimore had, indeed, 'evident cause' to rejoice at the triumph which had been achieved. All the important provisions of the bill, looked to the interests and had been framed with a view to the aggrandizement of the city. Its great leading object was, to secure the completion of the rail road to the Ohio river, and the completion of the canal to Cumberland, and its connexion with Baltimore by the route that might be found most conducive to the prosperity of that city. The enthusiasm of the occasion was, therefore, all embracing, on the part of the citizens of Baltimore. In the public demonstrations that were ordered, no dis-

[62] The canal had been built at this time only to Holman's Dam, twenty-six miles above Harper's Ferry, eighty-six miles from Washington; twenty-six miles more were under way.—*Report of the President and Directors. . . April 22, 1835.*

crimination was indicated, in regard to any particular work. No thought of jealous rivalry — no dream of future disappointment, or difficulty, was allowed to mingle in their exaltation at the auspicious event. But the act was not welcomed, by the Chesapeake and Ohio Canal Company, with the same satisfaction and pleasure. Indeed, by many of the stockholders, it was looked upon coldly, and, by some, positively objected to. Serious doubts were entertained, for a time, whether it would be accepted by the company." On July 28, 1836, however, the stockholders assented and agreed to the provisions of the Eight Billion Dollar Act.[63]

The directors of the Baltimore and Ohio Rail Road gave their assent to the law of 1835 on July 25, 1836. In addition to granting them the right to build their road up the Potomac Valley, the law allowed the city of Baltimore to subscribe to the stock, and, accordingly, Baltimore subscribed immediately for three million dollars worth of stock. Therefore within a year the assets of the road were increased by six million dollars.

[63]*Report of 1851*, p. 66.

Brighter days were now dawning for the road. The past six years had been a time of trials, drawbacks, and discouragements. The contest for a right of way to Harper's Ferry had been exasperating and had at last been won only by agreeing to limit the extension of the road to that point. There were other difficulties to be overcome before the new company could claim the genuine confidence of the public. All features of the road, excepting the road-bed alone, were experiments — rails, sleepers (ties), and cars. The road was opened May 22, 1830, and soon the public had passed a favorable verdict on the enterprise. In this day we would call the affair a horse-car railway. The only difference between this and other ordinary roads was the fact that the coach wheels ran on rails, being held in position by means of flanges. The coaches used were almost precisely like those on an ordinary pike, but were mounted on four light cast-iron wheels. Among roads — dirt, macadamized, plank, and cor-duroy — this road with rails was " the latest." As to its general practicability there was much discussion. What grades

could it overcome? Would curves be permitted as the scheme developed? As to its
popularity, no question could be raised.
Though the company had few cars and the
track was a single track and the road but
twelve miles long (running from Baltimore
to Ellicott's Mills), during the first four
months of operation the receipts were $20,-
012.36, and ten times the freight that could
be handled was offered.[64] An advertise-
ment of the rail road of 1830 is interesting.[65]
" Brigades " (trains) of cars left Baltimore
at 6 and 10 A. M. and from 3 to 4 P. M.;
brigades left the opposite terminus at " 6
and 8½ o'clock, A. M." and "12½ and 6
o'clock P. M." Drivers were not allowed
to permit passengers to enter the cars without tickets. A postscript reads: " P. S.
Parties desirous to engage a Car for the day
can be accommodated after the 5th July."

The question of motive power was the
great question of the hour. Horses and
mules only had been used on the other
two rail roads in Pennsylvania and Massachusetts; in this year (1830) on the Liver-

[64] Smith's *History and Description*, p. 25.
[65] Baltimore *American*, July 17, 1830.

pool and Manchester Rail Road steam
locomotives were used more successfully
than had been the case on other English
roads, where their speed had never exceeded
the gait of an easy-going road horse — six
miles an hour. It was greatly doubted
whether such a machine was possible; and
if, under good conditions, steam locomotives could haul a " brigade of cars " faster
than a horse or mule on straight track, the
thing would never get around a curve; and
it was never the plan of the builders of the
Baltimore and Ohio Rail Road to avoid
curves. Other locomotives than steam
were being prepared for trial on the new
rail road. Evan Thomas, brother of the
president of the road, invented a car which
was moved by sails! It was named
" Æolus." " I well recollect," recorded
Benjamin H. Latrobe, " the little experi-
mental locomotive of Mr. Evan Thomas; it
was ' a basket body,' like that of a sleigh,
and had a mast, and, if I recollect, ' a square
sail, and was mounted upon four wheels of
equal size.' It ran equally well in either
direction, but of course only in that in
which the wind happened to be blowing at

the time, although it would go with the
wind abaft the beam, but at a speed pro-
portioned to the angle with the plane of
the sails.   It was but a clever toy, but had
its use at the time in showing how little
power of propulsion was necessary upon a
railway, compared with the best of the
roads that had preceded it."[66]   The
" Æolus" attracted much attention; Baron
Krudener, envoy from the emperor of
Russia, made an excursion in the sailing
car, managing the sail himself.   On his
return he declared he had never before
travelled so agreeably, and remarked that
he 'would send his suite from Washington
to enjoy sailing on the Rail Road.'   The
President of the Company, to whom he had
been introduced, caused a model sailing car
to be constructed, fitted with Winans' fric-
tion wheels, which he presented to him,
with the reports that had been published
by the Company, to be forwarded to the
Emperor.   As a result Ross Winans of Baltimore was invited to Russia to take charge
of the emperor's plan of binding that empire
with railways.   His success marked one of

[66] Scharf's *History of Maryland*, vol. iii, p. 167.

the earliest if not the most spectacular instances of the success of American genius abroad." [67]

A horse-power locomotive was another invention, prior in date to the sailing car. " A horse was placed in a car and made to walk on an endless apron or belt, and to communicate motion to the wheels, as in the horse-power machines of the present day. The machine worked indifferently well; but, on one occasion, when drawing a car filled with editors and other representatives of the press, it ran into a cow, and the passengers, having been tilted out and rolled down an embankment, were naturally enough unanimous in condemning the contrivance. And so the horse-power car, after countless bad jokes had been perpetrated on the cowed editors, passed out of existence, and probably out of mind." [68]

The fate of the railway hung suspended on the successful solving of the question of motive power.[69] Peter Cooper's locomotive

[67] Smith's *History and Description*, pp. 25–27; Brown's *History of the First Locomotives in America*, p. 124. The name is here given as " Meteor."

[68] *Id.*, pp. 124–125.

[69] Peter Cooper to Wm. H. Brown, *Id.*, p. 109.

"Tom Thumb," constructed in 1829, at
Baltimore, and sent over the Baltimore and
Ohio Rail Road to Ellicott's Mills in one
hour and twelve minutes, August 28, 1830,
settled the momentous question.[70]   In
spite of its laughable features the picture
representing the " Exciting Trial of Speed
between Mr. Peter Cooper's Locomotive
' Tom Thumb,' and one of Stockton &
Stokes's Horse-Cars,"[71] in which the little
model locomotive has caught up with and
is passing the horse-car, represents nothing
less than the dawning of a new epoch in
human history.   Though improvements
were not made with great rapidity, they
came as fast as the rail road was able to
profit by them.   The Baltimore and Ohio
Rail Road merited the honorable title that
has been given it — the Railway University
of America.   While its rival, the Canal
Company, had a struggle to secure funds
to do its work, the railway carried the same
burden and with it the heavier burden of
doubt as to the future and many physical
and mechanical perplexities forever holding

[70]*Id.*, pp. 114–116.
[71]*Id.*, p. 119.

back successful realization of its schemes.
For illustration, take the question of track:
" The granite and iron rail; the wood and
iron on stone blocks; the wood and iron on
wooden sleepers, supported by broken
stone; the same supported by longitudinal
ground-sills, worked to a surface on one
side to receive the iron, and supported by
wooden sleepers; and the wrought iron
rails of the English mode; had all been laid
down, and as early as the year 1832, formed
different portions of the work." [72] With
the advent of the locomotive the light coach
wheels were replaced by cast-iron wheels
" to the perfection of which Ross Winans,
John Elgar, Jonathan Knight, and Phineas
Davis all contributed." [73] In 1832, steel
springs were placed upon a new locomotive " York " — built at York, Pennsylvania — and soon springs were placed on all
engines and cars. The discovery of the
advantage of combined cylindrical and

[72] Smith's *History and Description*, p. 33.

[73] Reizenstein's *Economic History*, p. 34. It is interesting to find Jonathan Knight, formerly Superintendent of the Cumberland Road in Ohio, now Chief Engineer of the Baltimore and Ohio Rail Road.—Cf. *Historic Highways of America*, vol. x, p. 91.

conical car wheels was a great forward
step helping to solve the question of turning curves sharply.   As early as 1831 the
Rail Road Company offered a prize of
$4,000 for the best locomotive offered for
trial on the road.[74]   The " York " was the
only engine of three offered that was capable of any good service.   Up to June 1834,
this engine, with the " Atlantic " and
" Franklin " were the only locomotives on
the road.   Horse-cars were still in common
use.   By the fall of 1834, five more locomotives were added and eight more had been
ordered.

Having passed through its darkest days
of struggle with the Canal Company and
with the vexatious problems of internal
betterment of rolling stock and motive
power, the Baltimore and Ohio Rail Road
was now in 1836, quite ready to take advantage of the provisions of the new law which
made it possible to throw its gleaming rails
up the Potomac from Harper's Ferry to
Cumberland and on to the coveted Ohio
Basin.   With the momentous question represented by the locomotive once solved

[74] Smith's *History and Description*, p. 30.

and solved forever, with an open route from tide-water to Cumberland and the West — little wonder that the controllers of the canal had been only lukewarm in their attitude to the Eight Billion Dollar Act! Despite their efforts, the railway was winning its way; with every new invention the West was made nearer the East; the locomotive was solving Washington's old question how the Potomac Valley could hold the West in fee. As Fate would have it — or Fortune — the hard labor and the thousand perplexities of many men from Washington down, who had attempted first to get in commercial touch with the West by means of rivers, then by means of a canal, were being swept aside by one blast of that little locomotive's whistle. How changed now the situation. But a few years back the canal was master of the Potomac Valley; it had allowed the feeble rail road a passage-way through the Point of Rocks only on condition that not one foot of track should be laid above Harper's Ferry until the canal had been completed to Cumberland. Now the canal was to receive sufficient state backing to complete its line

to Cumberland, on condition that the rail
road be allowed equal rights between
Harper's Ferry and Cumberland!   The
gloomy year of 1837 in the financial world
held the rail road back, and it was not
until 1839 that the work was actively
pushed on.   From now on there was no
delay; in June, 1842 the road was completed
to opposite Hancock, and by the end of the
year it was completed to Cumberland — one
hundred and seventy-eight miles from
Baltimore.   Exciting as is the story of the
westward movement of this giant, it cannot be treated here.   The first division to
Piedmont was opened in June 1851, not far
from the " blind " trace Washington rode
through far back in 1784, in search of a
portage road from eastern to western
waters.   By June, 1852, the road was
opened to Fairmont on the Monongahela,
and on the following January the first train
passed from Fairmont to Wheeling on the
Ohio.[75]   On the night of January 12, 1853
the banquet was spread in Wheeling to
end the day of celebration.   And of the
five " regular " toasts none was so typical

[75] Smith's *History and Description*, pp. 78–81.

or welcomed so loudly as that to " Thomas Swann:[76] Standing upon the banks of the Ohio, and looking back upon the mighty peaks of the Alleghanies, surmounted by his efforts, he can proudly exclaim — ' Veni, vidi, vici.' "

At the meeting of the Maryland legislature in December, 1838, the Chesapeake and Ohio Canal Company asked further assistance from the state and submitted an estimate of the work yet to be done to finish the canal to Cumberland. This estimate had been prepared by the chief engineer of the Chesapeake and Ohio Canal Company and reported to the board of president and directors January 22, 1839. Since the estimate of January, 1836, this was the first revised estimate, regarding quantities and including the extent of the whole line from Dam No. 5 to Cumberland, that had been made. Including a dam at the great Cacapon — now known as Dam No. 6 — but excluding the dam designated

[76] Mr. Swann was elected president of the rail road in 1848 and had ably conducted its affairs during the past five critical years, a worthy successor of Thomas and McLane.—*Id.*, p. 156.

as Dam No. 7, which was then temporarily
dispensed with, the estimate submitted ran
as follows:

For the 50 miles above the mouth of the Cacapon, now better known as Dam No. 6	$4,440,657.00
For the 27½ miles between Dam No. 5, and that point	1,640,000.00
	$6,080,657.00

Of this work there had been done, on the
first of December, 1838: $947,394.27 on the
50 miles; and $1,589,453.44 on the 27½
miles.   This left $3,543,809.29 as the work
remaining to be done on December 1,
1838, to complete the canal to Cumberland.
The work done in December was estimated
at about $90,000 which reduced the amount
remaining to be executed, on January 1,
1839, to about $3,450,000.   The twenty-
seven and a half miles between Dam No. 5
and Dam No. 6 were nearly completed at
the time the estimate was submitted.   In
April, 1839, navigation opened to Dam No.
6, which remained the western terminus
for a decade.

The $3,560,619 estimate of the seventy-

eight miles between Dam No. 5 and Cumberland, made in January 1836, was
arrived at by adding together the surveys
of two distinct parties of engineers. By
comparing the estimate submitted in December 1838 with the foregoing, made in
January 1836, which included the same
distance, it will be observed that although
the work on more than one-third of the
distance had been completed, the latter
estimate was nearly seventy-one per cent
in excess of the former. About fifty-seven
per cent of the increase was attributed by
the chief engineer to the advance in the
cost of labor, which was very high. The
pecuniary difficulties of the company, the
high prices and great difficulty in procuring
provisions along the line of the canal, and
the want of proper control over the
laborers by the civil authorities of the
state, were some of the causes contributing
to this excess. The remaining fourteen
per cent of the increase was stated to be
chargeable, mainly, to an increase of quantities found to be necessary in the progress
of construction for the security of the
canal. The " revised estimate " of Janu-

ary, 1839, was the last estimate upon which an available appropriation has been made to the Chesapeake and Ohio Canal Company by the state of Maryland.

A committee was appointed, after the presentation of the memorials and the revised estimate, to investigate the affairs and transactions of the company. In their report they expressed their belief in the importance of an early completion of the canal and suggested the expediency of an appeal to the general government. Instead of an appropriation by the state they recommended that a proposition be made to Congress, that the general government should either aid the company, or transfer to the state of Maryland the interest of the United States in its capital stock both as an original stockholder and as assignee of the district cities, on the condition that the state would provide the necessary means to complete the canal to Cumberland. A similar proposition had previously been made under joint resolutions adopted at December session, 1837, but nothing definite had resulted. The legislature, therefore, was not disposed to postpone the

advantages that were anticipated to result
from the completion of the work by a hope-
less recurrence of abortive expedients.
They were of the opinion that the state had
already gone too far in its investments in
the company to stop now — and it could
not recede. The unexpended and unen-
cumbered balance on hand was $681,853.59.
This was not sufficient to continue the
work during the present year. The credit
of the state was high and above suspicion.
Both applications were granted. An act
was passed on April 5, 1839, known as the
act of December session, 1838,[77] releasing
the Chesapeake and Ohio Canal Company
from the twenty per cent premium, stipu-
lated in the act of 1835, and authorizing
the commissioner of loans to issue to the
company five per cent sterling bonds to the
amount of $3,200,000 as an equivalent for,
and in lieu of, the $2,500,000 of six per
cent certificates which had been delivered
to the company, and the $500,000 of six per
cents which had been retained by the
treasurer of Maryland as security for the
payment of the premium. This act made

[77]*Maryland Laws*, 1838, ch. 386.

a further appropriation [78] which authorized an additional subscription to the capital stock of the company to the amount of $1,375,000 payable in five per cent sterling bonds.

These acts were promptly accepted by the Chesapeake and Ohio Canal Company and their provisions carried into effect and complied with. An instrument of guaranty, and mortgages to secure the payment of the three years interest, in compliance with the condition of both acts, were duly executed on May 15, 1839, and delivered to the treasurer of Maryland. The subscription of $1,375,000 authorized by the latter act was the last subscription made to the capital stock of the company. A report of the treasurer, issued June 1, 1839, stated that the means of the company, over and above its liabilities and applicable to the construction of the canal and the payment to the state of the interest on the bonds, amounted to $2,087,139.94. In this statement the whole amount of the sterling bonds was computed at par value. The cost of the remaining work to be done to

[78] *Maryland Laws*, 1838, ch. 396.

complete the canal, on the basis of the
January 1839 estimate, was, at that time,
$2,935,103.

At the following session of the general
assembly, December 1839, the company
made a formal application to the state for
further assistance. The accompanying
communication, dated February 10, 1840,
affirmed the correctness of the engineer's
estimate of January, 1839, and stated that
the fifty miles of canal between Dam No.
6 and Cumberland would cost $4,440,350.
Of this, $2,030,128 was expended on the
first of January 1840, leaving $2,410,222,
necessary to complete the work. The resources of the company, on the same day,
estimating 318,175 Maryland five per cent
sterling bonds at par, were stated to be
$1,489,571; the liabilities of the company
$1,244,555, leaving, January 1, after paying
all debts, a balance of $245,016. Upon this
exhibit, presented to the legislature, the
additional appropriation was asked for.

At this time the public appeared fully
cognizant of the great importance of press-
ing forward an early completion of the
canal. The members of the legislature

were also generally inclined to the adop·
tion of adequate measures of relief; but
the question which arose now was concerning the manner in which the relief should
be given.   Two ways were open: the state
bonds could be placed in the hands of the
commissioner of loans and be sold at par,
and the proceeds paid over to the Chesapeake and Ohio Canal Company; or, the
bonds could be delivered to the president
and directors of the company and sold by
them at par, or be exchanged àt their
nominal value for the evidences of debt of
the company.   Apparently, there was no
substantial difference between these two
propositions, but, because of the views and
feelings that originated and entered into
the controversy, a broad line of distinction
was drawn between the two plans.   Each
had its advocates and the supporters of
each were equally immovable — consequently the legislature adjourned without
making any appropriation at all.   In this
emergency the company took into consideration the course most proper to be
adopted in regard to continuing the
work on the canal.   When called upon to

present his views in reference to a total suspension of operations and the postponement of the completion of the canal, the chief engineer estimated that the accumulation of interest and other losses would amount to not less than a million dollars. Petitions from the contractors, merchants, and others, residing in the neighborhood of the operations, were received by the company, begging the continuation of the work and an issue of scrip, or promissory notes, which would be a convenience to each community. Accordingly the company decided to allow the work to proceed and to gratify the petitioners by issuing scrip. In 1839, and previous to that time, the issues had generally been secured by a pledge of state bonds or stocks. The present issue, on the other hand, which, during the year 1840, and from January to April 1841, amounted in the aggregate to $555,400, had no pledge to sustain it. It was the company's last issue of scrip.

At the December meeting of the Maryland legislature, 1840, an appropriation for aid was again asked for. The expenditures upon labor performed during the year had amounted to $531,160, and the sum required to finish the canal to Cumberland, according to the estimate of January 1839, was stated to be $1,825,892. In addition to this, estimating the unsold state bonds at eighty per cent, the company would need $700,000, exclusive of the interest due the state, to redeem the scrip and other debts. A committee was appointed by the legislature which made a rigid examination into the affairs and transactions of the company. The disposal of the state bonds and the issues of scrip were severely censured — and the general assembly again adjourned without adopting any measure of relief. Because of the threatening aspect of affairs, and the difficulty of procuring the necessary means for the continuance of the work on the canal, in the year 1839 the company began to cut down operations. In the month of May of that year the amount expended on the work was $96,320. In December, 1839, and January and February, 1840, the expenditure had been reduced to an average of $40,817 per month. The policy was also adopted of paying off the old loans, which had been secured by a pledge of the six per cent certificates without restriction as to sales, by an immediate sale of the five per cent sterling bonds. At a meeting of the stockholders of the company, on April 3, 1841, an adjourned session of a general meeting, the proceedings state that the president announced " if a breach should take place in the canal, the cost of repairing which might be $1500 or $2000, the money, and credit of the company, would not be sufficient to secure the repair of it, but that the company must, thereupon, be declared to be utterly bankrupt."

At an extra session of the Maryland legislature which was convened in March, 1841, by the proclamation of the governor, to provide means to pay the interest on the state debt, an application for further aid was again made by the company. On the fifth of April, 1841, an act was passed for an additional loan of two millions of dollars, payable in six per cent stock, or bonds of the state, which the legislature required to be sold by the treasurer of the state in behalf of the company. " The bonds were made to rest, upon the faith of the State and upon a specific pledge of the proceeds of the State's investments in the capital stock of the company, for the payment of principal and interest. The act, however, contained, as conditions precedent, clauses requiring the several companies of Alle-gany county, to enter into bond, satisfac-tory to the treasurer of the State, for the construction of a rail road, from the [coal] mines, to connect with the canal, and to complete the same simultaneously with its completion to Cumberland; and also, to guaranty the payment, to the company, of at least $200,000, per annum, for the trans-portation of their own coal on the canal.'' The board of directors, as well as the coal and iron companies of Allegheny County, made strenuous efforts to comply with these conditions, but the securities offered were not satisfactory to the treasurer of the state and the act failed. Later it was repealed.

At the December session, 1841, the pro-fessional beggar again asked aid of the legislature, but failed to secure it. For some time previous several contractors had been prosecuting the work on the canal on

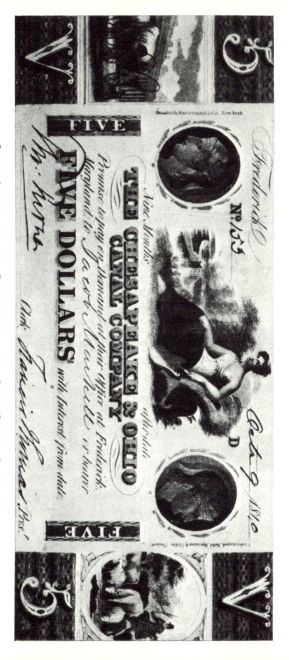

SCRIP ISSUED BY THE CHESAPEAKE AND OHIO CANAL COMPANY

their own credit and at their individual
expense, quite sure that the state would
make an appropriation at this session.
Failing to receive the desired assistance,
the work on the canal stopped abruptly
immediately after the adjournment of the
legislature, and before the end of the year
1841 not a man was in employ between Dam
No. 6 and Cumberland. At this time,
prostrate and overwhelmed with difficulties,
the company experienced great depression.
Not only were there great liabilities to the
state, secured by mortgage liens on the
canal and its revenues, but in addition to
this, the debts and obligations of the company due to individuals on scrip, accept-
ances, certificates of debt, common loans
and open amounts, as stated in the treasurer's abstract, on October 1, 1843,
amounted to $1,174,566.31. The urgent
appeals for payment coming from those
creditors, to whom large amounts were
due for work done, and who had been
quite reduced to poverty, excited general
sympathy. The canal had been completed
to Dam No. 6 in 1839, to which point it
was now only navigable. The chief engi-

neer estimated that it would cost $1,545,000 to complete the eighteen and three-tenths miles to Cumberland. The United States, the state of Virginia, the cities of the District of Columbia and all the stock-holders had long since discontinued their patronage and refused all pecuniary aid; even the state of Maryland, which had heretofore sustained the company and loyally upheld it in all its misfortunes, was now unable to give further assistance. The state was struggling under the evils of disordered finances and prostrate credit and a black shadow had been cast upon the name of the state because of its great debt contracted in behalf of the Chesapeake and Ohio Canal. Unable, because of the want of timely legislation, after the default of the internal improvement companies was made known, to meet her own public liabilities, she was certainly unable to give assistance to others. So the canal had no friend and no resources. The freshets of April and September, 1843, made heavy breaches in the canal which had to be repaired. This was done by the aid of accommodations from the banks and as a consequence the deficit was large and embarrassing at the close of that year. The entire revenues of the year only amounted to $47,635.51 and the current expenses to $83,792.80, showing a deficit of $36,157.29.

At the December session, 1844, application was renewed for a waiver of the state liens on the revenues of the canal so as to empower the company to issue its bonds, with preferred liens on its revenues to an amount not exceeding two millions of dollars. In principle and amount it was similar to the measure which had been proposed and rejected at the sessions of 1841 and 1842. Those who had been friends to the company during previous periods of difficulty were now conspicuous for their absence only, and the officers alone stood in vindication of the measure. Instead of a state convention and primary meetings to sustain and encourage the company, it was surrounded by enemies who opposed. The city of Baltimore took decided grounds in opposition to it, and the newspapers of the city were full of communications adverse to the proposed measure. The rail road company with diplomatic skill

sought to crush the effort by statements to
the effect that a connection between the
rail road and canal at Dam No. 6 would
render further prosecution of the canal
unnecessary.   It also with probably a similar object in view stated that " many years
would elapse, before the demand for coal
would require more than 100,000 tons, in
any one year, whatever facilities or transportation may be afforded."   Had the same
opposition been brought forward in December of 1834 or 1835 the work on the canal
at that time would probably have been
stopped; for, even with the powerful support of the immediate friends of the internal improvement companies, and the influential backing of the city of Baltimore, the
appropriations of 1834 and 1835 were
obtained only after a prolonged struggle,
especially on the part of the canal company.   The question of 1844 was one of an
entirely different nature.   It was not a
question of internal improvement — not
whether the wealth in the mountains
should be added to the general aggregate
of the state's resources, but a question of
finance — a question of whether the mil-

lions which the state had invested in the Chesapeake and Ohio Canal Company, should be given up as irretrievably lost or an effort be made to make the investment productive. Maryland had already expended seven millions on the work and had never expected any return from it until after completion. Neither money nor the state's credit was now asked, only that, since she herself was in pecuniary difficulties arising mainly from her support of the company (and these investments would remain unproductive until the completion of the canal) the state would waive her unprofitable liens on the revenues to such an extent as would enable the company to finish the work upon a preferred pledge of its future income. Although the opposition was great and influential there were Marylanders in the house of delegates at the December session, 1844, who believed in the importance of the completion of the canal and whose judgment was earnestly enlisted in favor of the plan. After a prolonged struggle the act waiving the liens of the state,[79] under which the

[79]*Maryland State Laws*, 1844, ch. 281.

canal was completed, was passed — passed
on the last day of the session, March 10, by
the limitation of the constitution, and
received a majority of one vote in each
house of the assembly! And, even then,
it had been so modified that its most
prominent advocates pronounced it value-
less and felt disposed to abandon its support!

In the charter of the Chesapeake and
Ohio Canal Company, prior to the year
1844, there was no express power to borrow
money for the completion of the canal, and
its right to do so had been much ques-
tioned. Even the force and validity of the
mortgages which it had given the state to
secure the payment of the two million loan
were called in question. Also, the time
limited by the charter for the completion of
the canal to Cumberland expired in 1840,
and since that time the corporation had
existed merely by the suffrance of the
power which had created it. No steps had
been taken to procure amendments in
either of these points. In the belief that
the measure suggested by the company for
the completion of the canal must prevail,

and that it would prove ineffectual unless
these defects in the charter were remedied,
the board of president and directors, at the
session of 1843, transmitted a memorial to
the legislature of Virginia asking for the
passage of an act providing for these
amendments. It also asked that the powers
of the company be enlarged, in regard to
extending the canal by a slackwater improvement to the mouth of Savage River
whenever it seemed expedient. This
memorial was accompanied by a draft of a
bill which embraced the desired provisions
and contained a reservation as to the liens
of Maryland. The legislature of Virginia
promptly acted and, with unimportant
changes, passed the bill on January 20,
1844. The act provided for an extension
of the time for the completion of the canal
to Cumberland to the first of January, 1855;
and authority was conferred upon the
president and directors, or a majority of
them assembled, '' to borrow money,   .   .
to carry into effect the objects authorized
by the charter of the company, to issue
bonds or other evidences of such loans, and
to pledge the property and revenues of the

company, for the payment of the same, and the interest to accrue thereon, in such form, and to such extent, as they may deem expedient; with a proviso saving the prior rights or liens of the state of Maryland, under the mortgages which had been executed by the company, to this state, except in so far as they should be waived, deferred, or postponed, by the Maryland legislature.'' When the company's acceptance was sent to the state treasurer, the act went into effect. The mortgage, bearing date of January 8, 1846, and executed in favor of the state of Maryland, was placed in the hands of the treasurer by the canal company. The amendments to the charter were ratified by Congress about one month before the passage of the act waiving the liens for the completion of the canal.

Considering the general depreciation of American securities and Maryland's discredit at this period, and the small means allowed for the accomplishment of the ends proposed, a sale of bonds at par was entirely out of the question. The only practicable course for the company to follow was a

resort to a contract payable in bonds covering all the subjects necessary to be provided for; this course was adopted. The
board, with the approval of the Maryland
state agents, after advertising for proposals,
concluded a contract which was carefully
guarded in all its provisions, on the twenty-
fifth of September, 1845. For the consideration of $1,625,000 of the bonds to be
issued under and pursuant to the act of
1844,[80] the four contractors pledged themselves to commence the work within thirty
days and finish the canal to Cumberland
within two years, according to the estimate
of 1842; they were, also, " to pay to a
trustee, for the use of the company, in
twenty-one monthly instalments, an aggregate sum of $100,000 in money, to enable
the Board of President and Directors to
liquidate land claims, engineering, and
other incidental expenses — and to pay the
interest on the bonds to be issued under
the act, until, and including the half year's
interest that would fall due, after the work
had been finished."

Very soon after the contract was made,

[80]*Maryland State Laws*, 1844, ch. 281.

the contractors commenced work on the canal between Dam No. 6 and Cumberland. All went well until the legislature again met and adjourned without restoring the credit of the state, when, their private means being exhausted, once more the contractors were compelled to suspend operations about June 1, 1846. The chief engineer's last report, made before the suspension, shows that the work done under the contract, according to the revised estimate of August 1845, amounted to $55,384. In addition to all other misfortunes, during the years 1846 and 1847 a series of freshets occurred in this region, one following the other in rapid succession. The lower division of the canal was repeatedly damaged until this increase of expense became very embarrassing. Nor were they able to make these repairs without the aid of temporary loans obtained from the banks.

After the execution of the contract for the completion of the canal, two of the original contractors of the co-partnership withdrew and Thomas G. Harris, of Washington County, Maryland, went in with the

remaining two — James Hunter, of Virginia, and William B. Thompson, of the
District of Columbia. The new firm was
called "Hunter, Harris and Co." In November, 1847, the contract was greatly modified;
the time for the completion was extended,
specific changes in the plan of construction
were made, and certain portions of the
work were entirely dispensed with — all
this with a view to a saving of cost, which
was absolutely necessary. Under the new
contractors' management operations were
quickly resumed, but, prosecuted under
such constant embarrassment, again ceased
March 11, 1850. The contractors made
great sacrifices in their sales of the bonds,
and, although stimulated to perseverance
in the honest expectation of completing the
canal, they had previously abandoned all
hope of profit, and found the pressure too
great to continue. This suspension, however, lasted but a few days. Hunter,
Harris, and Company made an assignment
of their interest in the contract to two of
their agents and attorneys, for the benefit
of their creditors, and the work was recom-
menced and continued until July, 1850,

when it was again abandoned because of
the usual lack of means.   Upon the seventeenth of July the board of president and
directors formally declared the canal
abandoned and on the following day entered into a new contract with Michael
Byrne, of Frederick county, for the completion of the canal to Cumberland.   The
work remaining to be done was inconsid-
erable, yet tedious, consisting of numerous
unfinished portions between Dam No. 6
and Cumberland.   This work was promptly
commenced and diligently prosecuted, and
the canal was opened for navigation purposes, and through trade commenced, on
October 10, 1850.   Mr. Byrne continued to
press forward the work, which did not
interfere with the passage of the boats, and
on February 17, 1851, the final payment
was made to him under the provisions of
the contract.   From this time the completion of the Chesapeake and Ohio Canal may
be dated.

From the clerk's statement made from
the books of the company, with an additional allowance for a few small unsettled
claims, it appears that " the cost of the

Chesapeake and Ohio Canal, from the mouth of the Tyber in the city of Washington, to the town of Cumberland, a distance of one hundred and eighty-five and seven tenths miles, for construction, engineer expenses, lands, and other contingencies properly applicable to construction, amounts, in the aggregate, to the sum of $11,071,176.21, or $59,618.61 per mile." It is interesting to note that the original estimate for a canal of less dimensions, made by the experienced General Bernard in 1826, was $8,177,081.05, or $43,963 per mile. This estimate did not embrace land purchases or condemnations nor make any provision for contingencies with the exception of an allowance of $157,161 for fencing, which, in the statement of cost, is included under the head of lands. Therefore, in order to make a just comparison between the original estimate and the actual cost of the canal there should be added to General Bernard's estimate the cost of those items which are excluded from it, and included in the clerk's statement, after deducting the amount for fencing already embraced in the estimate.

Bernard's estimate, exclusive of land purchases, condemnations and contingencies $8,177,081.05

Add the items excluded, viz., actual cost of lands, deducting therefrom $157,161, for fencing, already embraced in the estimate 267,562.91

Engineer expenses 429,845.94

Incidental damages 28,870.09

Pay of officers, say 80,000.00

Total $8,983,359.99

Aggregate actual cost, as per clerk's statement . 11,071,176.21

Excess of actual cost over original estimate, with the above additions — twenty-three and one-fifth per cent, or $2,087,816.22

It is rather a difficult undertaking to give a brief yet succinct and accurate history of the old waterway since 1850. In a nutshell, the history of the Chesapeake and Ohio Canal from its completion to 1889 may truthfully be said to be a history of the Democratic party in the state of Maryland during that period. It was used as a political machine and lever by that party at the expense of its physical and financial good. The officers of the company were appointed by the Board of Public Works of the state, some of whom were ex officio members of the board of directors. The members of the Board of Public Works were appointed by the governor of the state, and in that way the management of the canal was controlled by the party in power, which, during that period, was the Democratic party. There was much litigation in an effort by some of the holders of bonds to protect themselves, but it was always unsuccessful. Mr. Gorman, now Senator A. P. Gorman, was president for a number of years. It is an open secret throughout the state that it was on the placid waters of the Chesapeake and Ohio Canal that the

senator rode into the high dignity of a
Senatorial seat. The canal was in every
way a financial failure and paid nothing to
the holders of its debentures. There are
today thousands of dollars of unpaid wages,
due for labor and material supplied. It has
cost the state of Maryland millions of dollars, none of which are likely to ever find
their way back to the state coffers. Conducted upon an economical and business-
like basis, it should have been a source of
revenue.

The disastrous floods of 1889 caused
such damage to the waterway that a large
sum was required to restore it. The state
refused further financial aid and, in consequence, the canal lay abandoned. The
Democratic politicians of the state, many
of whom were interested in the West
Virginia Central Railway, made an effort,
through an act passed in Maryland legislature, to sell the valuable property and its
franchises to that rail road for a nominal
price; in fact were on the point of dispos-
ing, for about two hundred thousand dollars, of a property worth millions. After the
passage of the act and its signature by the

A VIEW OF THE CHESAPEAKE AND OHIO CANAL

[This part of the canal, at the entrance of the tunnel thirty miles east of Cumberland, shows the expensive nature of portions of the work]

governor, the holders of the bonds which were authorized to be issued in 1844, and which were issued in 1848, stepped in. When these bonds were authorized there were already so many liens upon the canal that it was a well-known fact that no market would be found for them. Realizing this fact, the state, to give them a value, waived its rights, under previous issues and loans, as we have seen, in favor of these bonds about to be put upon the market, and also securing them by a mortgage on the tolls and revenues of the canal. The holders of these bonds arose and petitioned the courts to protect them, claiming that a sale of the canal to the rail road would destroy the corpus, and that with the corpus destroyed, the toll and revenue earning capacity would cease. In other words they claimed that a mortgage on the tolls and revenues constituted a mortgage on the corpus. They further petitioned that the court appoint trustees to operate the canal for the bondholders of 1848, thereby enabling them to have an opportunity to protect themselves. The case was bitterly fought in the courts and

ended finally by the granting of the peti-
tion.   Trustees were appointed for a term
of four years to show what they could do.
Then the canal was repaired, at a cost of
over half a million dollars.   In 1891
traffic was resumed and has been going
steadily on since that time.   That the
court is evidently satisfied with the showing made by the trustees is attested by the
fact that, at the expiration of the four years
originally granted (in which to show that
they could run the canal successfully) the
court granted an extension of that time for
four years more, and at the expiration of
the latter grant, further increased it four
years, and so on.

The Canal is now, as it has been since
1891, operated by the trustees, under mortgage of Chesapeake and Ohio Canal Company, dated June 5, 1848.

CHAPTER IV

THE PENNSYLVANIA CANAL AND ITS SUCCESSOR

THE great early trade route through Pennsylvania in the days of the pack-horse and " Conestoga " wagon has been outlined in previous volumes of this series.[81] By means of Forbes's Road the metropolis of the United States at the beginning of the nineteenth century, Philadelphia, was in close connection with the metropolis of the Ohio Basin, Pittsburg. The rivalry with Baltimore had been keen, and the Philadelphia merchants were eager to overcome their handicap of nearly one hundred miles, by internal improvements of a most advanced pattern. In the matter of roads, liberal as had been Pennsylvania's policy, Maryland was far ahead, so far as the West was concerned. And in 1806, when a national road across

[81] *Historic Highways of America*, vols. v, xi, xii.

the Alleghenies was proposed, and a Maryland (Cumberland) route was chosen by the commissioners appointed by President Jefferson, it seemed probable that Maryland's lead in the matter of trade was about to be materially increased.

But Pennsylvania, as we have seen, had been an early promoter of inland navigation; its " Society for promoting the improvement of roads and inland navigation" in 1791, had called specific attention to the rivers which should be made important routes of an expanding commerce. Among the most important recommendations of this society was that looking to the improvement of Pennsylvania's great western waterway, the Susquehanna River and its tributary, the Juniata. This latter stream interlocked, beyond the Allegheny crest, with the roaring Conemaugh, a tributary of the Kiskiminitas and Ohio. And in response to this appeal we have seen that £5,250 was appropriated to the improvement of Susquehanna navigation from Wright's Ferry to the mouth of Swatara Creek. As Philadelphia was the commercial center, the route thence by water was first up the Schuylkill, then across by canal to the Susquehanna. It was outlined as follows in the society's memorial, signed by Robert Morris, February 7, 1791:

	Miles	Chains
Up Schuylkill to the mouth of Tulpehocken	61	00
Thence up Tulpehocken to the end of the proposed canal	37	09
Length of the canal	4	15
Down Quitipahilla to Swatara	15	20
Down Swatara to Susquehanna	23	00
Up Susquehanna to Juniata	23	28
Up Juniata to Huntingdon	86	12
From Huntingdon, on Juniata, to the mouth of Popular run.	42	00
Portage to the Canoe Place on the Conemaugh	18	00
Down Conemaugh to Old Town at the mouth of Stoney Creek	18	00
Down Conemaugh and Kiskeminetas to Allegheny	69	00
Down Allegheny river to Pittsburgh on the Ohio	29	00
Total	426	04[89]

The progress of Virginia and Maryland in connection with the Potomac Company and the opening of the Potomac River was felt in Pennsylvania at this time. " For,

[89] *An Historical Account* . . *of Canal Navigation in Pennsylvania* (Philadelphia, 1795), p. 3.

in the firſt place," read this memorial,
" if we turn our view to the immenſe territories connected with the Ohio and Mif-
ſiſſippi waters, and bordering on the great
lakes, it will appear from the tables of
diſtances, that our communication with
thoſe vaſt countries (conſidering Fort Pitt
as the port of entrance upon them) is as
eaſy and may be rendered as cheap, as to
any other port on the Atlantic tide waters.
The diſtance from Philadelphia to the
Allegheny, at the mouth of Kiſkeminetas,
is nearly the ſame as from the mouth of
Monongahela to George Town on Potomac;
and ſuppoſing the computed diſtances from
Pittſburgh to the Dunkard Bottom to be
juſt, and the navigation of Cheat river, on
the one hand, and the Potomack, at the
mouth of Savage river, on the other, to be,
at all feaſons of the year, equal to the navigation of the Kiſkeminetas, Conemaugh
and Juniata; yet as the portage from
Dunkard Bottom to the Potomack, at the
mouth of Savage river, is thirty-ſeven
miles and a quarter, and the portage from
Conemaugh to Juniata only eighteen miles
(which may be conſiderably ſhortened by

locks) there can be no doubt but that the tranſportation of all kinds of goods and merchandize from Philadelphia to Pittſburgh may be at a much cheaper rate than from any other ſea port on the Atlantic waters.''

It mattered not where it was, every one of the Atlantic seaboard cities had an expert who could show in black and white that that particular port was in closest touch with Pittsburg and the West. Washington had done so, conclusively to all Southerners; Morris does it here to the satisfaction of Pennsylvanians, and New York had a score of mathematicians who could prove the same thing concerning New York, and the Hudson and Mohawk route.

'' This is not mentioned,'' continues the memorial, hopefully, '' with a view to diſparage the internal navigation of our ſiſter ſtates, more eſpecially *Maryland* and *Virginia*. We admire their noble exertions.   .   .   But, although a conſiderable part of the ſettlers on the Ohio waters may be accommodated by the Potomack navigation, and the ſtate of Pennſylvania

may only have a fhare in the trade of thofe
waters; yet there remains to us the im-
menfe trade of the lakes, taking Prefqu'Ifle,
which is within our own ftate, as the great
mart or place of embarkation." [83]

It is exceedingly interesting to note that
while Pennsylvania at this time only
expected to share with her southern neighbors the trade of the Ohio Basin, she
expected a monopoly of the trade on the
Great Lakes. Of the latter trade she
secured only a fraction, while of the former
she secured practically a monopoly for half
a century.

The route is more carefully outlined in
the memorial: " It connects Philadelphia
with Pittfburgh and all the Ohio waters, by
the Schuylkill, the Swatara and Juniata
branches of Sufquehanna, and the Kifkeminetas branch of Allegheny, with the
diftance of five hundred and fixty-one miles
and an half   .   .   and alfo Philadelphia
and Prefqu'Ifle, ufing the fame waters   .   .
to the mouth of Kifkeminetas, and then by
the eafy waters of Allegheny and French
Creek.   In this whole communication to

[83] *Id.*, pp. 7-8.

Pittsfburgh, there are only eighteen miles portage between the Juniata and Cone-maugh . . and only the addition of fifteen miles and an half more at the port-age from Le Bœuf to Prefqu'Ifle, which portage is, likewife, included in both the other communications. In this ftatement of portages, it is fuppofed that the canal or lock navigation between the heads of Tulpehocken and Quitipahilla, is to be com-pleated; but if that work fhould be thought too great to begin with, it will be only the addition of four miles portage, by an excellent and level road.''

For many years the problem of the navigation of this westward waterway was the subject of discussion and legislation. In no case does any state seem to have profited by the experience of any other. New York, Pennsylvania, Virginia, and Maryland each boldly attacked the problem of the improvement of their rivers, the Mohawk, Juniata, and Potomac, without in the least profiting by the experience of the others. It was New York which first broke away from the old ideas, upon which millions of dollars had been squandered,

and built her Erie Canal, which soon
turned doubt and derision into a vast
tumult of applause. Yet it must be re-
membered that the New York canal had
an easy path to follow. The Mohawk was
not the wild Potomac as known in bleak
Hampshire County, nor Wood Creek the
racing Conemaugh or upper Youghiogheny. The wet flats of the " Genesee
Country " offered a different prospect for
canal engineers from that to be viewed in
Kittanning Gorge where only the eagles
lived. The Erie Canal conquered, by
means of locks, 500 feet in 360 miles; the
Chesapeake and Ohio Canal faced the
problem of overcoming 2754 feet in 340
miles, and the Pennsylvania Canal, 2291
feet in 320 miles. It is not to be won-
dered at, then, that New York found a
water connection with the West first. And
yet the fact remains that much was spent
on the Mohawk before the Erie Canal was
begun.

So the struggle went on in Pennsylvania
for nearly a generation until at last the success of the Erie Canal and the failure of
the improved unnavigable rivers gave birth

to the Pennsylvania Canal.[84]   On March 27, 1824, an act of the Pennsylvania legislature authorized the appointment of a board of canal commissioners to view and explore routes for a canal from Harrisburg to Pittsburg.   The commissioners, Colonel Jacob Holgate, James Clark, and Charles Treziyulney, were appointed March 31.   From May until December they were in the field. Their exploration resulted in the following estimate of the height to be overcome between the Susquehanna[85] and Ohio:

	Rise (feet)
Harrisburg (mouth of Juniata) to head of Juniata	589
Head of Juniata to proposed tunnel	945
Tunnel level to summit of mountain	754
Susquehanna to mountain summit	2288

[84] One of the most enlightening broadsides of the time treating of the delay of the internal improvement plan is Turner Camac's *Facts and Arguments respecting the great Utility of an extensive plan of Inland Navigation in America* (Philadelphia, 1805).

[85] The canal was proposed to begin on the Schuylkill and lead to the Susquehanna, but it actually began on the Susquehanna, the country between that point and Philadelphia being covered by the old Union Canal and the Columbia Railway which was soon built.

The tunnel at summit level was as long as that one proposed on the Chesapeake and Ohio Canal, and gave rise to great discussion. One commissioner, Treziyulney, whose name gave weight to his opinion, disagreed with his associates on the matter of the tunnel, and, in fact, on the entire canal proposition. The majority report having been made to the governor of Pennsylvania February 2, 1825, this minority report was dated February 21. " In short," it read, " the whole country, from the upper forks of the Juniata to the forks of the South branch of the Conemaugh, is mountainous; mountain rising after mountain in quick succession. The main one where the proposed tunnel is to pass, is hemmed in and surrounded by other high mountains, with steep slopes separated from one another by narrow ravines and presenting no favorable situation for canaling, either by lockage or tunneling. Here nature has refused to make her usual kind advances to aid the exertions of man; mountains are thrown together, as if to defy human ingenuity, and baffle the skill of the engineer." The difference of opinion caused much debate and conjecture as to the practicability of the great plan.

Another cause of delay was the agitation which was now sweeping all thinking minds on the question of the new roads of (literally) iron and steam as a motive power. Such had been the progress of railways in England that it was believed by many that this method of locomotion would supersede all others. On February 5, 1825, the Pennsylvania senate granted the wish of the advocates of railways by appointing a commission to inquire into the possibility of building a railway from Philadelphia to Pittsburg. Three editions of a pamphlet, *Facts and Arguments in favor of adopting railways in preference to Canals in the State of Pennsylvania*, were published in Philadelphia this year. It maintained that a railway could be built from Philadelphia to Pittsburg in one-third the time it would take to build a canal and at one-third the cost; it would, moreover, be available almost the entire year, whereas the Erie Canal was navigable only two hundred and twenty days in the year. It urged that more persons would be required in the operation of a canal than a railway, and that the tolls would be higher on the former than on the latter. "If a railway, or even a canal, existed between Pittsburg and Philadelphia, New Orleans would not requite the consideration of a moment. The great distance of this port from Kentucky, Indiana, Ohio, and Pennsylvania; in winter the ice in the Ohio river . . the numerous sawyers, snags &c. . . the length of the voyage . . are powerful objections to this port. . . Baltimore presents itself as the second rival. . . But when the Pennsylvania railway shall be constructed, Baltimore cannot for a moment withstand the competition of the enormous capital of Philadelphia. She may, indeed, construct a canal or a railway . . but little is to be apprehended, as the length and expense of constructing these works will be far greater than those contemplated in Pennsylvania. New York is the third rival . . but the communication between New York and Pittsburg must be effected by a long, tedious, and expensive

voyage, *requiring four changes of ves-
sels.   .   . The route of nearly 800 miles*
[via Buffalo, Lake Erie, and the Allegheny
River] *will be very circuitous; and will be
impracticable five months every year.   .   .*
It does not require the voice of prophecy
to predict *that the period is not far distant
when the New York canal will be superseded by
a railway.*"[86]

Thus was the question of rivalry between
the Atlantic ports stated by a railway
exponent of 1825; and the statements must
be considered extremely prophetic.   The
railway commission was appointed too far
ahead of the times, but it had a forward
influence, and by the act of April 11, the
canal commissioners were authorized to
have all routes across the Alleghenies surveyed, and in June of the year following
the Juniata route was announced to be the
preferable route in the commission's report
to the governor of Pennsylvania; the tunnel, however, was considered impossible
for the same reason as with the tunnel
between the heads of Savage and Youghiogheny Rivers in western Virginia — the

[86] *Facts and Arguments*, pp. 57-58.

difficulty of supplying water at the summit level. In the place of a tunnel, inclined planes were proposed by the commission.

Evidently anticipating this report, the Pennsylvania legislature passed an act for the construction of the Pennsylvania Canal at the expense of the state; and it was approved by Governor Shulze, February 25, 1826. It read:

AN ACT TO PROVIDE FOR THE COMMENCE-
MENT OF A CANAL, TO BE CONSTRUCTED
AT THE EXPENSE OF THE STATE, AND TO
BE STYLED THE PENNSYLVANIA CANAL.

Whereas, The construction of a canal within our own limits, for the purpose of connecting the eastern and western waters, is believed to be practicable, and within the means of the state, and its speedy completion will advance the prosperity, and elevate the character of Pennsylvania, and by facilitating intercourse, and promoting social interests, will strengthen the bands of the Union; *And whereas*, There are important sections of the work which may be immediately begun without the danger of error:

Therefore,

Section 1. *Be it enacted by the Senate and House of Representatives of the Commonwealth of Pennsylvania, in General Assembly met,* That the commissioners appointed by the act, entitled, "An act to appoint a board of canal commissioners," passed the 11th April, 1825, are hereby authorized and empowered, in behalf of this state, immediately to locate and contract for making a canal, and locks, and other works necessary thereto, from the river Swatara to the mouth of the river Juniata, and also from Pittsburg to the mouth of the Kiskiminetas.

Section 2. *And be it further enacted by the authority aforesaid,* That the said commissioners shall be authorized to appoint one or two of the board, as occasion may require, as acting commissioner or commissioners, who shall receive —— dollars per day, while actually engaged in the superintendence of the works contemplated by this act.

Section 3. *And be it further enacted by the authority aforesaid,* That the said board shall appoint a treasurer, and shall have power to appoint engineers, clerks, and other officers, toll-gatherers, and such other agents as they shall judge requisite, and to agree for, and settle their respective wages, and to establish reasonable toll. *Provided*, That the treasurer shall give bond in such penalty, and with such security, as the said board shall direct, for the true and faithful discharge of the trust reposed in him.

Section 4. *And be it further enacted by the authority aforesaid*, That the location and dimensions of the said canals and locks shall be determined by a majority of the board, with the approbation of a skilful engineer, and with the consent of the Governor.

Section 5. *And be it further enacted by the authority aforesaid*, That it shall and may be lawful for the said board, or a majority of them, to agree with the owners of any land, through which the said canal is intended to pass, for the purchase, use and occupation thereof, on behalf of the state, and in case of disagreement, or in case the owner thereof shall be a feme covert, under age, non compos, or out of the state or county, on application to a justice of the

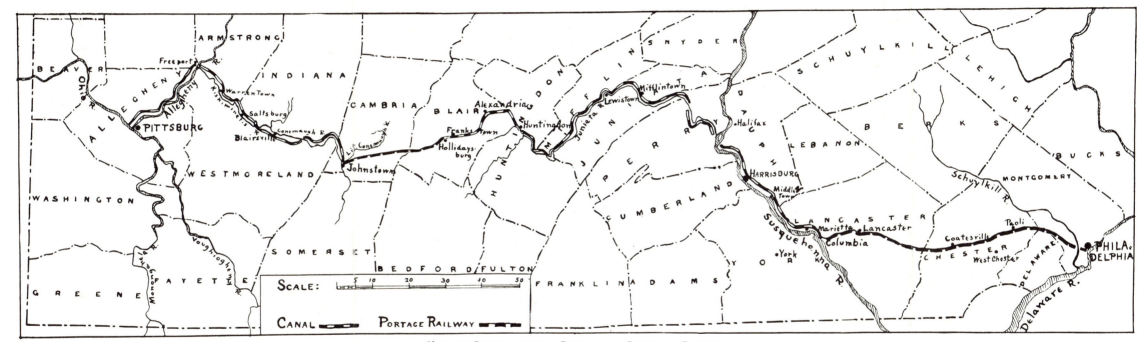

MAP OF PENNSYLVANIA CANAL AND PORTAGE RAILWAY

SCALE: 5 10 20 30 40 50

CANAL ⊏⊐⊏⊐   PORTAGE RAILWAY ▬▬▬

ARMSTRONG

BEAVER

Freeport   R.

INDIANA

Warren Town

Kiskiminetas R.

Saltsburg

ALLEGHENY

OHIO R.

PITTSBURG

CAMBRIA   BLAIR

Alexandria

Blairsville

Conemaugh R.

L. Conemaugh R.

Franks Town

Huntingdon

Hollidays-burg

Johnstown

WESTMORELAND

WASHINGTON

SOMERSET

FAYETTE

Youghiogheny R.

Monongahela R.

West Chester

GREENE

BEDFORD FULTON

SNYDER

MIFFLIN

Mifflintown

Lewistown

Juniata R.

HUNTINGDON

JUNIATA

PERRY

Halifax

Harrisburg

Middle Town

CUMBERLAND

FRANKLIN ADAMS

YORK

York R.

Susquehanna R.

SCHUYLKILL

LEHIGH

LEBANON

BERKS

BUCKS

Schuylkill R.

MONTGOMERY

LANCASTER

Marietta   Lancaster

Columbia

Coatesville

CHESTER

West Chester

Paoli

PHILA. DELPHIA

Delaware R.

county in which such land shall be, the
said justice of the peace shall issue his
warrant, under his hand, to the sheriff of
the county, to summon a jury of eighteen
inhabitants of his county, not related to
the parties, or in any manner interested,
to meet on the land to be valued, at a day
to be expressed in the warrant, not less
than ten nor more than twenty days thereafter; and the sheriff, upon receiving the
said warrant, shall forthwith summon the
said jury, and when met, shall administer
an oath or affirmation to every juryman
who shall appear, being not less than
twelve in number, that he will faithfully,
justly and impartially value the land, and
all damages the owner shall sustain by
cutting the canal through such land, or the
partial or temporary appropriation, use or
occupation of such land, according to the
best of his skill and judgment, and that in
such valuation he will not spare any person, for favor or affection, or any person
grieve for malice, hatred or ill will; and in
every such valuation and assessment of
damages, the jury shall be, and they are
hereby instructed to consider in determin-

ing and fixing the amount thereof, the actual benefit which will accrue to the owner, from conducting the said canal through, or erecting any of the said works upon his land, and to regulate their verdict thereby, except that no assessment shall require any such owner to pay or contribute any thing where such benefit shall exceed, in the estimate of the jury, the value and damages ascertained as aforesaid; and the inquisition thereupon taken, shall be signed by the sheriff, and some twelve or more of the jury, and returned by the sheriff to the clerk or prothonotary of his county, and unless good cause be shown against the said inquisition, it shall be affirmed by the court and recorded; but if the said inquisition should be set aside, or if from any cause, no inquisition shall be returned to such court within a reasonable time, the said court may at its discretion, as often as may be necessary, direct another inquisition to be taken in the manner above described, and upon every such valuation, the jury is hereby directed to describe and ascertain the bounds of the land by them valued, and the quality and duration of the interest and estate in the same, required by the said board for the use of the state, and their valuation shall be conclusive on all persons, and shall be paid for by the said board, to the owner of the land, or his legal representatives; and on payment thereof, the state shall be seized of such lands, as of an absolute estate in perpetuity, or with such less quantity and duration of interest or estate in the same, or subject to such partial or temporary appropriation, use or occupation as shall be required and described as aforesaid, as if conveyed by the owner, and whenever, in the construction of the said canal, or any of the works thereof, locks, dams, ponds, feeders, tunnels, aqueducts, culverts, bridges or works of any other description whatsoever appurtenant thereto, it shall be necessary to use earth, timber, stone or gravel, or any other material to be found on any of the lands adjacent or near thereto, and the said board or their agent cannot procure the same for the works aforesaid, by private contract of the proprietor or owner, or in case the owner should be a feme covert, non compos, or under age, or out of the state or county, the same proceedings in all respects shall be had as in the case before mentioned, of the assessment and condemnation of the lands required for the said canal, or the works appurtenant thereto.

Section 6. *And be it further enacted by the authority aforesaid*, That every person actually engaged in labouring on any canal authorized by law, shall be exempt from doing militia duty in this state, except in cases of insurrection or invasion, during the time when he is so actually engaged; and the certificates of the contractor who shall employ such men, so liable to perform militia duty, in the performance of their contracts shall be *prima facia* evidence of such engagement.

Section 7. *And be it further enacted by the authority aforesaid*, That the sum of three hundred thousand dollars be, and the same is hereby appropriated, to be paid by the state treasurer, in such sums as shall be required for the execution of the work, which sums shall from time to time be paid into the hands of the treasurer of the board by direction of a majority of the board, and by warrant of the Governor.

The method of joining the two divisions of what now became known as the Pennsylvania Canal was left undecided, pending further investigation.   But an act of March 24, 1828 authorized the location, and construction of the Juniata division, from Lewistown to the highest practicable point on the river.   Eventually Hollidaysburg on the eastern slope of the Alleghenies, and Johnstown, on the western, were decided upon as the termini of the eastern and western divisions of the canal.   The thirty-six miles intervening were to be crossed by a railway, through Blair's Gap, of the inclined planes previously suggested.

The interminable delays and postponements which have been described on the Chesapeake and Ohio Canal were unknown in the present instance.   What was known as the central division (there being really no eastern, though the Union Canal was such nominally) was begun at Columbia on the eastern shore of the Susquehanna, July

4, 1826, and was opened to Duncan's Island, above Harrisburg, in 1830. This central section also extended across the Susquehanna and up the Juniata Valley; it was begun in 1827 and completed to Huntingdon in 1830, and Hollidaysburg in 1834. The western division of the canal extended from Pittsburg to Johnstown; it was begun in 1826 and opened in 1830.[87]

From Huntingdon on the east to Johnstown on the west of the mountains was planned the Allegheny Portage Railroad, which from any point of view must be considered one of the most interesting and remarkable of all attempts to abridge distance in our early history.

The history of the divisions of the Pennsylvania Canal on either side of the mountains is commonplace beside this interesting and daring bit of engineering. Inclined planes were not, at this time, a novelty, but their use as proposed now in the Alleghenies was

[87] For this and many additional items of information concerning the greater problem of Pennsylvania's entire system of canals, see Theodore B. Klein's monograph " The Canals of Pennsylvania and the System of Internal Improvements," *Report Pennsylvania Secretary of Internal Improvements, 1900.*

such an advance on former instances that
it was considered a bold experiment.   The
Morris Canal between the Hudson and
Delaware " is peculiar," wrote the British
engineer Stevenson, in 1837, " as being
the only canal in America in which the
boats are moved from different levels by
means of inclined planes instead of locks.
The whole rise and fall on the Morris Canal
is 1557 feet, of which 223 feet are overcome by locks, and the remaining 1334 feet
by means of twenty-three inclined planes,
having an average lift of 58 feet each.   .   .
The car [on which canal boats ascend and
descend]   .   .   consists of a strongly
made wooden crib or cradle   .   .   on
which the boat rests, supported on two iron
waggons running on four wheels.   When
the car is wholly supported on the inclined
plane, or is resting on the level, the four
axles of the waggons are all in the same
plane   .   .   ; but when one of the
wagons rests on the inclined plane, and the
other on the level surface, their axles no
longer remain in the same plane, and their
change of position produces a tendency to
rock the cradle, and the boat which it sup-

ports; but this has been guarded against in
the construction of the boat-cars on the Morris Canal by introducing two axles .   .
on which the whole weight of the crib and
boat are supported, and on which the
waggons turn as a centre.   The cars run on
plate rails laid on the inclined planes, and
are raised and lowered by means of machinery driven by water wheels.   .   .
The railway, on which the car runs, ex-
tends along the bottom of the canal for a
short distance from the lower extremity of
the plane; when a boat is to be raised, the
car is lowered into the water, and the boat
being floated over it, is made fast to the
part of the framework which projects above
the gunwale.   .   .   The machinery is
then put in motion; and the car bearing
the boat, is drawn by a chain to the top of
the inclined plane, at which there is a lock
for its reception." [88]

The building of such inclined planes on
the Allegheny Portage Railway from Hollidaysburg to Johnstown marks the first
conquest of that thousand mile summit

[88] *Sketch of the Civil Engineering of North America*,
pp. 128–129.

line of the Alleghenies.   For fifty years,
in the infancy of engineering, American
promoters had been struggling with this
problem ; it is a far cry — measured by the
hosts of futile plans and dreams — from
Washington, pushing his horse through
the dripping laurels along " McCullough's
Path," to Sylvester Welch who spanned
Blair's Gap by a railway; then, and not
until then, was a passage-way from the
Atlantic seaboard to the Mississippi Basin
open for freight and passengers on which
neither freighter nor coach played any
part.   The building of the Allegheny
Portage Railway, 1830–33, was as epoch-
making an event as the opening of the
Cumberland Road in 1818 or the opening
of the great trans-Allegheny railways at
the middle of the century.   In many ways
it was more significant than the opening of
the Erie Canal, which was merely a lengthy
application of a principle already perfectly
understood.   Considering the coach and
wagon to have been natural means of communication, we can then say that the Portage Railway was the first artificial means
of communication between the East and

the Mississippi Basin. Well did Mr. Stevenson say that in boldness of design and in difficulty of execution this railway could be compared with no modern work he had seen, unless exception be made for the passes of the Simplon and Mount Cenis in Sardinia; and these, as engineering works, did not impress him as more wonderful.

The project had been proposed early in the history of the canal and in 1826 the experienced Erie Canal engineer Canvass White delivered an opinion that the plan was feasible, but added that a portage wagon road would perhaps answer temporary needs. As the canal building advanced in the valleys on either side of the mountains, the plan of a connecting link which would satisfactorily mount the towering crest which intervened was seriously debated. Late in 1828 Moncure Robinson became engineer in charge and went into the field in 1829 with plans well developed; in November he reported to the board of canal commissioners that the crest could best be overcome by a system of inclined planes, with stationary engines; near the

summit, a tunnel a mile in length was
planned to pierce the crest one hundred
and seventy-seven feet beneath its summit
and twelve hundred and sixty-four feet
above Hollidaysburg, the starting point of
the inclines.   The total cost for a railway
thirty miles in length, of the pattern described, was estimated at slightly less than
a million dollars ($936,004.87).   On June
8, 1830 a board of engineers consisting of
Robinson, Lieutenant-colonel S. H. Long,
and Major John Wilson was appointed to
survey the route proposed and make final
recommendations.   Late in that year a
report was made which conformed largely
with Mr. Robinson's plan matured in 1829,
and on March 21, 1831, Governor Wolf
approved " An act to continue the improvement of the State by canals and railroads."
Section 3 of that act read:

" *And be it further enacted by the authority
aforesaid*, That the said canal commissioners
shall commence forthwith and prosecute
without delay, a rail road over and across
the Allegheny mountains, from the basin at
Hollidaysburg, in the county of Huntingdon, to Johnstown, in the county of Cam-

bria, a distance of about thirty eight
miles, according to the extent, route and
plan thereof, stated in their report of the
twenty-first day of December, one thousand eight hundred and thirty, excluding
the plan of a tunnel as recommended by
Moncure Robinson in his report of the
twenty-first November, one thousand eight
hundred and twenty-nine; and also, that
they shall commence and prosecute, without delay, the extension of the Juniata
division of the Pennsylvania canal, from
the town of Huntingdon, in the county of
Huntingdon, to the basin at Hollidaysburg, in the same county, either by canal
or slack water navigation; towards the
expenditures of which rail-road and canal
or slack water navigation, as specified in
this section, during the present year, the
sum of seven hundred thousand dollars is
hereby specifically appropriated, to be
paid out of the loan hereinafter mentioned." [89]

The great work was now actually begun.
Sylvester Welch, formerly superintendent of the western division of the canal,

[89]*Laws of Pennsylvania*, 1830–31, no. 104 (p. 195).

was made principal engineer, and Mr.
Robinson consulting engineer; Samuel
Jones was superintendent. The final
surveys were conducted from Johnstown
to the mountain summit beginning in
April; they were completed by May 20,
1831, and the work let to the lowest bid-
ders at Ebensburg May 25. The surveys
on the eastern slope of the mountain were
conducted from Hollidaysburg and were
completed in the July following. The
contracts were let at Hollidaysburg on
July 29.[90]

The termini of the road were at the
canal basins at Hollidaysburg and at Johnstown, the former 1,398 feet below the
mountain summit, and the latter 1,771 feet
below the summit. The road occupied a
clean swath through the forests, of one
hundred and twenty feet in width, lest fall-

[90] These, as well as many preceding and succeeding
data, are from William Bender Wilson's admirable
monograph, " The Evolution, Decadence, and Abandonment of the Allegheny Portage Railroad " in the *Annual
Report of the Secretary of Internal Affairs of the
Commonwealth of Pennsylvania*, 1898–99, part iv, pp.
xli–xcvi. This monograph forms an important chapter
in Mr. Wilson's *History of the Pennsylvania Railroad
Company*.

ing trees should damage the work. All surveys, estimates and recommendations to the contrary notwithstanding, Mr. Robinson's tunnel was not now built, another plane being added to the total which enabled the railway to vault the summit. The planes were ten in number; beginning at Johnstown they were as follows:

				Length (feet)	Elevation (feet)
Plane No.	1	.	.	. 1,607.74	150.00
	2	.	.	. 1,760.43	132.40
	3	.	.	. 1,480.25	130.50
	4	.	.	. 2,194.93	187.86
	5	.	.	. 2,628.60	201.64
	6	.	.	. 2,713.85	266.50
	7	.	.	. 2,655.01	260.50
	8	.	.	. 3,116.92	307.60
	9	.	.	. 2,720.80	189.50
	10	.	.	. 2,295.61	180.52

There were six levels between the planes on the western division and five on the eastern. The level between Johnstown and the foot of Plane No. 1 was four miles in length, and that between Planes 1 and 2 was thirteen miles in length, overcoming an elevation of 189.58 feet. The

THE FIRST AMERICAN TUNNEL

[By this tunnel the Allegheny Portage Railway crossed the summit of the mountains]

remainder on that division were about a
mile in length, each rising about twenty
feet. The shortest level on the eastern
slope was .15 of a mile and the longest 3.72
miles, descending 146.71 feet. The steepest
incline rose only 10¼ feet in a hundred —
a grade not much steeper than that on
many pioneer roads. Mr. Welch affirmed
that cars could be drawn up these by
horses or by stationary steam or horse-
power engines. On the eastern planes he
suggested that advantage be taken of the
force of gravity; on three of the levels,
those at each terminus, and the thirteen
mile level between Planes 1 and 2, he
urged the use of locomotives; elsewhere he
advised the use of horse power. The road
(single track) was completed by the beginning of 1834 and traffic began March 18,
1834. The ten planes were supplied with
ten stationary engines. Half a century
ago, Washington, in that classic appeal to
Harrison, of 1784, maintained that a great
plan of communication between the East
and West was practicable, in that " The
western inhabitants would do their part
towards its execution. Weak as they are,"

he said, "they would meet us half way."
What a splendid comment it is on Washington's wisdom and foresight to record
that these engines on the Allegheny Portage Railway, which hauled the first load
of freight over the Alleghenies which ever
crossed them by artificial means were made
in the young West — in Pittsburg! Washington, at least, did not misjudge in the
least the spirit of those Virginians and
Pennsylvanians who in his day were pushing ahead over Indian trails into the lands
beyond the mountains.

The second track of the railway was put
under contract at Hollidaysburg May 31,
1834. In the same year three locomotives
for the levels were ordered, one from Bos-
ton, and two from Newcastle, Delaware.
One of these was sent on to Pittsburg by
canal to serve as model of others to be built
there. "The road as completed," writes
Mr. Wilson, "showed a width of track
between rails of 4 feet and 9 inches, and a
distance between tracks, including width
of inner rail of each track, of 5 feet. The
railway between the planes was laid to
correspond vertically with the grade

adopted for the road, and was in all cases laid to form horizontal arcs of circles, or their tangents. Flat iron bars on wooden rails were placed on the inclined planes. On the balance of the road, edge rails 18 feet in length, weighing 39½ pounds to the yard were laid, resting in iron chairs on wooden sills. The latter were fastened to cross ties where the road passed over high embankments, but, on solid ground they were attached to stone blocks measuring about 3½ cubic feet. To do this two holes were drilled into each block. Into these holes oak plugs were driven. The cast-iron chair was placed directly upon the top of the stone block, and spikes driven through holes in the flanges of the chair into the oak plugs. The rail was a double headed rail, and held in place by a wedge. The difficulty of the spreading of the tracks was at first overcome by substituting for each alternate pair of blocks a stone block some 7 feet long, extending across the track, and having a chair at each end. This was found to be too expensive, and wooden cross ties were placed between each pair of stone blocks.''

The road as opened was, like the original Baltimore and Ohio Rail Road, merely a new sort of road-way, on which horses drew cars on rails (instead of on a flat road-bed) between the inclines. A rush of business at once overwhelmed the road. Between the middle of March and the middle of April, 1834, the number of cars tripled in number and were then entirely inadequate to the trade. Much "portaging" was done in the old way on the old-time portage path by wagon. The business was done by transportation firms or by individuals, the commonwealth furnishing the road-bed, and a motive power only on the inclined planes.

It was in October, 1834, that the keel-boat "Hit or Miss" from the Lackawanna, Jesse Crisman owner and Major C. Williams commander, first of all craft to leap the Alleghenies, was taken from Susquehanna waters at Hollidaysburg and laid safely in Allegheny waters at Johnstown. Crisman expected to sell his boat at Hollidaysburg — as his ancestors had ever done; but John Dougherty of the Reliance Transportation Line, constructed a car calculated to bear " the novel burden." Starting at noon, "they rested at night on the top of the mountain, like Noah's Ark on Ararat, and descended the next morning into the Valley of the Mississippi, and sailed for St. Louis." [91] It was fifty years, to the month, since the pioneer promoter of trans-Allegheny communications, Washington, was searching in Dunkard Bottom for a pathway for keel-boats across this great divide. History was again repeated; as in the old days when, in 1758, Forbes's Road through Pennsylvania eclipsed Virginia's highway which Washington championed, Braddock's Road, because it was a more direct route from the heart of colonial life to the Ohio Basin, so now Pennsylvania's waterways, joined by a portage railway of only thirty-eight miles in length, eclipsed any and all other possible water routes to the Ohio Valley by being actually opened to commerce. These repetitions of history illustrate Pennsylvania's keystone position in the United States, so far as the seaboard and the commercial centers

[91] Sherman Day's *Historical Collections of the State of Pennsylvania*, p. 184.

of the Mississippi Basin are concerned.

Though exceptionally interesting and suggestive, the Allegheny Portage Railway was only a link in a chain. The two great canals, in the valleys of the Juniata and Conemaugh were the greater links; a horse-car rail road was laid from Philadelphia to Columbia on the Susquehanna and this, soon supplied with locomotives, became the eastern link in the chain of communication of which the Pennsylvania Canal was the important part. By 1835 the complete system was in operation between Philadelphia and Pittsburg; a table of distances will be interesting:[92]

THE PENNSYLVANIA ROUTE IN 1835

Division No. 1

Columbia Rail Road

	Miles from Philadelphia
Fair Mount Water Works . .	1
Viaduct over Schuylkill River .	3

[92] H. S Tanner's *A Brief Description of the Canals and Rail Roads of the United States* (November, 1834), pp. 25–26.

Paoli	20½
Downingtown	32
Coatsville	40
Mine Ridge	52½
Lancaster	69½
Mt. Pleasant	76½
Columbia	81¾

Division No. 2

Central Division of the Pennsylvania Canal

	Miles from Philadelphia
Marietta	84¾
Bainbridge	91¼
Falmouth	94¾
Middletown	99
Harrisburg	108
Duncan's Island	124½
Newport	135
Mifflintown	157
Lewistown	171
Waynesburg	185
Aughwick Falls	197
Jack's Mt.	203
Huntingdon	214
Petersburg	221

Alexandria	.	.	.	.	. 228
Frankstown	.	.	.	.	. 250½
Hollidaysburg	.	.	.	.	. 253½

Division No. 3
Allegheny Portage Rail Road

			Miles from Philadelphia
Walker's Point	.	.	.	. 255
Inclined Plane No. 10	.	.	. 257¼
" " No. 6	.	.	. 263¾
Mountain Br.	.	.	.	. 272¾
Ebensburg Br.	.	.	.	. 275¾
Staple Bend	.	.	.	. 285¾
Johnstown	.	.	.	. 290¼

Division No. 4
Western Division of the Pennsylvania Canal

			Miles from Philadelphia
Laurel Hill	.	.	.	. 297
Lockport	.	.	.	. 307
Blairsville	.	.	.	. 320
Saltzburg	.	.	.	. 336
Warrenton	.	.	.	. 348
Leechburg	.	.	.	. 358
Aqueduct over Allegheny River	. 361
Freeport	.	.	.	. 363

Logan's Ferry 376
Pine Creek. 388
Pittsburg 394¼

The lockage in the central division between Columbia and Hollidaysburg was
747¾ feet; it was forty feet wide at the
top, twenty-eight feet at the bottom, and
was four feet deep.   The dams numbered
eighteen; there were thirty-three aqueducts and one hundred and one locks, including guards; those between Columbia
and Duncan's Island were 90 x 17 feet;
the remainder 90 x 15 feet.   About sixteen miles on the Juniata was slackwater navigation in 1834.   The western
division was the same in width and
depth as the central; the lockage from
Johnstown to Pittsburg was 471 feet.   On
this division there were sixty-four locks,
90 x 15 feet, ten dams, two tunnels, sixteen
aqueducts, sixty-four culverts, thirty-nine
water weirs and one hundred and fifty-two
bridges; 21¼ miles was slackwater navigation.   The cost of the central division
was $5,307,253.26; the Juniata Valley por-
tion costing $3,570,016.29.   The western
division cost $3,096,522.30; making the

entire original cost of the canal proper
$8,403,775.56. The total original cost of
the Allegheny Portage Railway to January,
1837, including laying the second track
and building the Conemaugh viaduct was
$1,634,357.69¾, making the total cost of the
"Pennsylvania Canal" $10,038,133.25¾ —
half a million dollars more than the Erie
Canal, which it also exceeded in length by
thirty-one miles. Yet the Chesapeake and
Ohio Canal, of only one hundred and
eighty-five miles in length, cost a million
dollars more than the Pennsylvania Canal.

The later history of the Pennsylvania
Canal well illustrates the restlessness of
human hearts, and the mighty conquests
over nature which restless ambition has
made possible. One success, such as the
Portage Railway, only suggested a greater
one, a railway over the mountain. The
road was only in fairly good working order
when, in 1836, the Pennsylvania legislature
passed a resolution ordering the canal commissioners to have a survey made of the
Alleghenies to determine whether the
inclined planes could not be dispensed
with! Within the next decade a New

Portage Railway was planned which would follow, in part, the route of the old line. The hundreds who were connected with the manipulation of the expensive and cumbersome planes decried, of course, the new road, as we have noted so often in this series; the owners and operators of earlier methods of transportation scoffed at and opposed the new. But the new rail road was not built at once. The opposition carried weight. In April, 1846, however, the Pennsylvania Railway was incorporated to build a through thoroughfare from Philadelphia to Pittsburg. Of all routes the Juniata-Conemaugh passage-way offered an unrivaled course and was quickly chosen. The long contest over right of way in the Potomac Valley could not be reproduced here, as the canal was a state affair. In 1847, contracts had been let for sections eastward from Pittsburg and westward from Harrisburg. In two years the sixty miles between Harrisburg and Lewistown were opened; in the year following the portion from Lewistown to Hollidaysburg was completed. The western division was pushed up the Conemaugh with equal

rapidity and on December 10, 1852, communication between the termini was possible, passengers and freight being trans-
ferred across the mountain crest by stage
and wagon.   In 1854 the railway was completed across the Alleghenies.

In 1850 the legislature took steps to improve the communication between the two
ends of the canal by building the proposed
portage road and avoiding planes.   The
work went on simultaneously with the
building of the Pennsylvania track; as a
temporary accommodation the railway company allowed the portage operators to avoid
Plane No. 1, by using the railway track for
a distance of four miles east from Conemaugh station, east of Johnstown.   Planes
No. 2 and No. 3 were avoided by means of
a new double track to the foot of Plane
No. 4.   In 1854 the Pennsylvania Railway
was completed across the mountain, and
the trade of that company was of course
lost to the Portage Railway.   On July 1,
1855, the new portage route was in operation, though incomplete.

The great success of the Pennsylvania
Railway and its importance to the commer-

cial interests of the commonwealth tended
to sink the old canal and its portage railway out of sight. In 1855 this "main line
of the public works" was offered for sale,
but the offer was not liberal. Another act
was passed May 16, 1857, for its sale and
June 25, it was purchased by the Pennsylvania Railway Company; possession was
taken in August.

After attempting to operate successfully
the Portage Railway, the new owners lost
$7,220.14 in three months and ordered the
line closed. The canal was operated by
the Pennsylvania Canal Company in the
interest of the railway and that was gradu-
ally abandoned. The division from Pittsburg to Johnstown was entirely abandoned
by 1864; the portion in the Juniata Valley
was abandoned in 1899, and that along the
Susquehanna in 1900.

Appendixes

APPENDIX A

AN ACT FOR OPENING AND EXTENDING THE
NAVIGATION OF POTOWMACK RIVER [93]

I. Whereas the extension of the navigation of Potowmack river, from tide water to
the highest place practicable on the North
branch, will be of great public utility, and
many persons are willing to subscribe
large sums of money to effect so laudable
and beneficial a work; and it is just and
proper that they, their heirs, and assigns,
should be empowered to receive reasonable
tolls forever, in satisfaction for the money
advanced by them in carrying the work
into execution, and the risk they run: And
whereas it may be necessary to cut canals
and erect locks and other works on both

[93] We present here the first three sections of the act
as given in Hening's *The Statutes at Large; being a
collection of all the Laws of Virginia from the first
session of the Legislature in the year 1619* .   .
(Richmond, 1823), vol. xi, 9th of Commonwealth, ch.
xliii, October, 1780.

sides of the river, and the legislatures of
Maryland and Virginia, impressed with
the importance of the object, are desirous
of encouraging so useful an undertaking:
Therefore,

II. *Be it enacted by the General Assembly
of Virginia*, That it shall and may be lawful to open books in the city of Richmond,
towns of Alexandria and Winchester in this
state, for receiving and entering subscriptions for the said undertaking, under the
management of Jaquelin Ambler and John
Beckley at the city of Richmond, of John
Fitzgerald and William Hartshorne at the
town of Alexandria, and of Joseph Holmes
and Edward Smith at the town of Winchester, and under the management of such
persons and at such places in Maryland as
have been appointed by the state of Maryland, which subscriptions shall be made
personally or by power of attorney, and
shall be paid in Spanish milled dollars,
but may be paid in foreign silver or gold
coin of the value; that the said books shall
be opened for receiving subscriptions on
the eighth day of February next, and continue open for this purpose until the tenth

day of May next, inclusive; and on the
seventeenth day of the said month of May,
there shall be a general meeting of the
subscribers at the town of Alexandria, of
which meeting notice shall be given by the
said managers, or any four of them, in the
Virginia and Maryland Gazettes, at least
one month next before the said meeting;
and such meeting shall and may be continued from day to day until the business
is finished; and the acting managers at the
time and place hereinafter mentioned,
shall lay before such of the subscribers as
shall meet according to the said notice, the
books by them respectively kept, containing the state of the said subscriptions; and
if one half of the capital sum hereinafter
mentioned, should, upon examination, appear not to have been subscribed, then the
said managers at the said meeting, are empowered to take and receive subscriptions
to make up the deficiency; and a just and
true list of all the subscribers, with the
sums subscribed by each, shall be made
out and returned by the said managers, or
any four or more of them, under their
hands, into the general court of each state,

to be there recorded; and in case more
than two hundred and twenty-two thousand two hundred and twenty-two dollars
and two ninths of a dollar, shall be subscribed, then the same shall be reduced to
that sum by the said managers, or a
majority of them, by beginning at and
striking off a share from the largest subscription or subscriptions, and continuing
to strike off a share from all subscriptions
under the largest, and above one share,
until the sum is reduced to the capital of
two hundred and twenty-two thousand two
hundred and twenty-two dollars and two-
ninths of a dollar, or until a share is taken
from all subscriptions above one share, and
lots shall be drawn between the subscribers
of equal sums, to determine the numbers
in which such subscribers shall stand, on
a list to be made for striking off as aforesaid; and if the sum subscribed still
exceeds the capital aforesaid, or all the
subscriptions are reduced to one share:
and if there still be an excess, then lots to
be drawn to determine the subscribers
who are to be excluded, to reduce the
subscriptions to the capital aforesaid,

which striking off shall be certified in the
list aforesaid, and the said capital sum
shall be reckoned and divided into five
hundred shares of four hundred and forty-
four dollars and four-ninths of a dollar
each, of which every person subscribing
may take and subscribe for one or more
whole shares, and not otherwise. *Provided*,
That unless one half of the said capital
shall be subscribed as aforesaid, all subscriptions made in consequence of this act,
shall be void, and in case one half and less
than the whole of the said capital shall be
subscribed as aforesaid, then the president
and directors are hereby empowered and
directed to take and receive the subscriptions which shall first be offered in whole
shares as aforesaid, until the deficiency
shall be made up, a certificate of which
additional subscriptions shall be made
under the hands of the president and
directors, or a majority of them for the
time being, and returned to and recorded
in the general courts, aforesaid.

III. *And be it enacted*, That in case one
half of the said capital, or a greater sum,
shall be subscribed as aforesaid, the said

subscribers, and their heirs and assigns, from the time of the said first meeting, shall be, and are hereby declared to be incorporated into a company, by the name of the "Potowmack Company," and may sue and be sued as such; and such of the said subscribers as shall be present at the said meeting, or a majority of them, are hereby empowered and required to elect a president and four directors, for conducting the said undertaking, and managing all the said company's business and concerns, for and during such time, not exceeding three years, as the said subscribers, or a majority of them, shall think fit. And in counting the votes of all general meetings of the said company, each member shall be allowed one vote for every share, as far as ten shares, and one vote for every five shares above ten, by him or her held at the time in the said company; and any proprietor, by writing under his or her hand, executed before two witnesses, may depute any other member or proprietor to vote and act as proxy for him or her, at any general meeting.

APPENDIX B

AN ACT INCORPORATING THE CHESAPEAKE AND OHIO CANAL COMPANY [94]

Whereas a navigable canal from the tide water of the river Potomac, in the District of Columbia, to the mouth of Savage creek, on the north branch of said river, and extending thence across the Alleghany mountain, to some convenient point on the navigable waters of the river Ohio, or some one of its tributary streams, to be fed, through its course on the east side of the mountain, by the river Potomac and the streams which empty therein, and on the western side of the mountain, and in passing over the same, by all such streams of water as may be beneficially drawn thereto by feeders, dams, or any other practicable mode, will be a work of great

[94] This act, of which we present here the first two sections, was passed by the Virginia legislature January 27, 1824. It was published in *Laws Relative to the Chesapeake and Ohio Canal* (Washington, 1827).

profit and advantage to the people of this
State, and of the neighboring States, and
may ultimately tend to establish a connected navigation between the eastern and
western waters, so as to extend and multi-
ply the means and facilities of internal
commerce, and personal intercourse between the two great sections of the United
States, and to interweave more closely all
the mutual interests and affections that are
calculated to consolidate and perpetuate
the vital principles of Union; and whereas
it is represented to this General Assembly
that the Potomac Company are willing and
desirous that a charter shall be granted to
a new company, upon the terms and conditions hereinafter expressed; and that the
charter of the present company shall cease
and determine:

1. *Be it therefore enacted by the General
Assembly of Virginia*, That, as soon as the
Legislatures of Maryland and Pennsylvania, and the Congress of the United
States, shall assent to the provisions of this
act, and the Potomac company shall have
signified their assent to the same, by their
corporate act, a copy whereof shall be

delivered to the Executives of the several States aforesaid, and to the Secretary of the Treasury of the United States, there shall be appointed by the said Executives and the President of the United States, three commissioners on the part of each State, and the Government of the United States, any one of whom shall be competent to act for his respective government. The said commissioners shall cause books to be opened at such times and places as they shall think fit, in their respective States, and the District of Columbia, under the management of such persons as they shall appoint, for receiving subscriptions to the capital stock of the company hereinafter incorporated; which subscriptions may be made, either in person or by power of attorney; and notice shall be given in such manner as may be deemed advisable, by one or more of the said commissioners, of the time and places of opening the books.

2. And the said commissioners shall cause the books to be kept open at least forty days. And within twenty days after the expiration thereof, shall call a general

meeting of the subscribers at the city of
Washington, of which meeting notice
shall be given, by a majority of the commissioners aforesaid, in at least four of the
newspapers printed in Pennsylvania, Maryland, Virginia, and the District of Columbia, at least twenty days next before the
said meeting; and such meeting shall, and
may be continued from day to day until
the business is finished; and the commissioners at the time and place aforesaid,
shall lay before such of the subscribers as
shall meet according to the said notice, the
book containing the state of the said subscriptions: and, if one fourth of the capital
sum of six millions of dollars should appear
not to have been subscribed, then the said
commissioners, or a majority of them, at
the said meeting, are empowered to take
and receive subscriptions to make up such
deficiency, and may continue to take and
receive such subscriptions for the term of
twelve months thereafter; and a just and
true list of all the subscribers, with the
sum subscribed by each, shall be made out,
and returned by the said commissioners, or
by a majority of them, under their hands,

to the Board of Public Works of this State,
to the Governor and Council of the State of
Maryland, to the Secretary of State of the
State of Pennsylvania, and to the Secretary
of the Treasury of the United States, to be
carefully preserved; and in case more than
six millions of dollars shall be subscribed,
then the sum subscribed shall be reduced
to that amount, by the said commissioners,
or a majority of them, by beginning at and
striking off a share from the largest subscription or subscriptions, and continuing
to strike off a share from all subscriptions
under the largest and above one share,
until the same is reduced to the capital
aforesaid, or until a share is taken from all
subscriptions above one share; and lots
shall be drawn between subscribers of
equal sums, to determine the number of
shares which each subscriber shall be
allowed to hold, on a list to be made for
striking off as aforesaid: and if the sum
subscribed still exceed the capital aforesaid,
then to strike off by the same rule, until
the sum subscribed is reduced to the capital aforesaid; or all the subscriptions reduced to one share respectively: and, if

there still be an excess, then lots shall be
drawn to determine the subscribers who
are to be excluded, in order to reduce the
subscription to the capital aforesaid; which
striking off shall be certified on the lists
aforesaid; and the said capital stock of the
company, hereby incorporated, shall con-
sist of six million of dollars, divided into
sixty thousand shares, of one hundred dollars each; of which every person subscribing may take, and subscribe for one or
more whole shares; and such subscriptions
may be paid and discharged either in the
legal currency of the United States, or in
the certificates of stock of the present
Potomac company, at the par or nominal
value thereof, or in the claims of the creditors of the said company, certified by the
acting president and directors to have been
due, for principal and debt, on the day on
which assent of the said company shall
have been signified by their corporate act
as hereinbefore required: *Provided*, that the
said certificates of stock shall not exceed,
in the whole amount, the sum of three
hundred and eleven thousand one hundred
and eleven dollars and eleven cents; nor

the said claims the sum of one hundred and
seventy-five thousand eight hundred dollars: *Provided, also*, that the stock so paid
for in certificates of the stock of the present
company, shall be entitled to dividend,
only as hereinafter provided: and that no
payment shall be received, in such certificates of stock, until the Potomac company
shall have executed the conveyance pre-
scribed by the thirteenth section of this
act: *And, provided*, that, unless one-fourth
of the said capital shall be subscribed, as
aforesaid, all subscriptions made in consequence of this act shall be void; and, in
case one fourth, and less than the whole
capital, shall be subscribed as aforesaid,
then the said commissioners, or a majority
of them, are hereby empowered and
directed to take and receive the subscriptions, which shall first be offered in whole
shares, as aforesaid, until the deficiency
shall be made up; a certificate of which
additional subscription shall be made,
under the hands of said commissioners, or
a majority of them, for the time being,
and returned as aforesaid.